Fundamentals of Metallurgical Thermodynamics

Santosh Kumar Sahoo · Mithilesh Kumar ·
Swapan Kumar Karak

Fundamentals of Metallurgical Thermodynamics

Springer

Santosh Kumar Sahoo
Department of Metallurgical and Materials Engineering
National Institute of Technology Rourkela
Rourkela, Odisha, India

Mithilesh Kumar
Department of Metallurgical and Materials Engineering
National Institute of Technology Rourkela
Rourkela, Odisha, India

Swapan Kumar Karak
Department of Metallurgical and Materials Engineering
National Institute of Technology Rourkela
Rourkela, Odisha, India

ISBN 978-981-99-6670-7 ISBN 978-981-99-6671-4 (eBook)
https://doi.org/10.1007/978-981-99-6671-4

This Springer imprint is published by the registered company Springer Nature Singapore Pte Ltd.
The registered company address is: 152 Beach Road, #21-01/04 Gateway East, Singapore 189721, Singapore

Preface

Thermodynamics is a basic subject and has found wide applications in mechanical, chemical, ceramics, metallurgical and materials engineering, etc. The subject "Metallurgical Thermodynamics or Thermodynamics of Materials" is taught to the students as a compulsory subject in their undergraduate and post-graduate programmes in all universities and institutes. Thermodynamics has been a confusing subject for the beginning students due to conceptual difficulties experienced by them. Many of the average students get terrified by this subject and take it mechanically for solving the numericals only. Students need to have a proper knowledge of this subject for understanding other subjects at higher levels in course curriculum of metallurgical and materials engineering. Nature and style of presentation of the book greatly affects in creating interest of the students for the subject. Basic concepts of the subject need to be explained in the textbook in easy manner and to a sufficient degree of clarity, particularly for the beginners. To make the subject graspable, convincing approach needs to be followed for enunciation and explanation of fundamental principles to promote their understanding. Although few books written by renowned thermodynamicists have been published so far, they are voluminous, non-exhaustive and not easily understandable by the beginning students and thus not suitable as introductory textbooks. Hence, there is an urgent need to publish more concise and easily understandable textbook of metallurgical thermodynamics in the market for considerable benefits of both the students and teachers.

Keeping in mind of the requirements of the students and on the basis of our teaching experience of this subject for around 27 years in undergraduate level, we have made a sincere attempt to illustrate the fundamentals and their applications with the help of appropriate examples of metallurgical interest. This textbook on *Fundamentals of Metallurgical Thermodynamics* includes almost all the important basic topics, derivations and numericals for graduating students of chemical, ceramics and metallurgical and materials engineering and is expected to be a reference book for them. In addition to these students, researchers and practicing engineers can refresh their knowledge in this subject. In general, metallurgical/materials thermodynamics is associated with chemical thermodynamics. In the present textbook having six chapters, thermodynamic properties in context to both the high- and low-temperature physico-chemical processes and solid/liquid solutions have been dealt with. Important thermodynamic features like ΔH, ΔS, ΔG, equilibrium calculations, etc. associated with the above processes have been discussed in detail. The laws of thermodynamics are general, and their applications depend upon the type of discipline. In total, around 60 problems have been solved and 35 given as an exercise in this book. Despite taking all the feasible care, some unidentified errors or confusion or miscalculation may be present in this textbook. Any advice from the readers for further refinement in this book will be deeply appreciated by the authors. The authors greatly acknowledge the help of Sristi Kumar, M.Sc. Student, Central University of Jharkhand, India, for her valuable aid in writing this book. We are also thankful to the publisher, Springer Nature, Switzerland, for its regular backing in preparation and publication of this book.

Rourkela, India

Santosh Kumar Sahoo
Mithilesh Kumar
Swapan Kumar Karak

Contents

1 Introduction 1
1.1 Introduction 1
1.1.1 Historical Perspective of Classical Thermodynamics 2
1.1.2 Limitations of Classical Thermodynamics 3
1.1.3 Thermodynamics and Energy Parameters 4
1.2 Thermodynamics Systems 4
1.3 State of a System 5
1.4 Extensive and Intensive Properties 6
1.5 Thermodynamic Processes 6
1.5.1 Reversible and Irreversible Processes 7
1.6 State and Path Functions 9
1.7 Thermodynamic Equilibrium 10
1.8 Concluding Remarks 12

2 First Law of Thermodynamics 15
2.1 Concept of Internal Energy 15
2.1.1 Internal Energy in Terms of Partial Derivatives 17
2.2 Statements of the First Law of Thermodynamics 18
2.2.1 Statements of the First Law of Thermodynamics for Different Systems 19
2.2.2 Implications of the First Law of Thermodynamics 19
2.3 Heat Capacity 22
2.3.1 Relation Between C_P and C_V 23
2.3.2 Dependence of Heat Capacity on Temperature 25
2.3.3 Importance of Heat Capacities C_P and C_V 26
2.4 Enthalpy 27
2.4.1 Enthalpy Changes 28
2.4.2 Classification of Enthalpy Changes 28
2.5 Thermochemical Laws 32
2.5.1 Hess's Law 32

2.5.2 Kirchhoff's Law ... 34
2.6 Internal Energy Versus Enthalpy ... 35

3 Second Law of Thermodynamics and Entropy ... 47
3.1 Introduction ... 47
3.2 Observations Leading to the Formulation of Second Law ... 48
3.3 Statements of Second Law of Thermodynamics ... 49
3.4 Mathematical Formulation of Second Law of Thermodynamics ... 50
3.4.1 Conversion of Heat to Work by a Reversible Cycle ... 51
3.4.2 Carnot Cycle with Ideal Gas as Working Substance ... 52
3.4.3 Generalized Carnot Cycle ... 55
3.5 Entropy ... 55
3.5.1 Entropy: A State Property ... 56
3.6 Entropy Changes in Reversible Processes ... 57
3.6.1 Reversible Isothermal Expansion of an Ideal Gas ... 58
3.6.2 Reversible Adiabatic Expansion of an Ideal Gas ... 59
3.6.3 Reversible Heating at Constant Volume or Variation of Entropy with Temperature Under Isochoric Condition ... 59
3.6.4 Reversible Heating at Constant Pressure or Variation of Entropy with Temperature Under Isobaric Condition ... 60
3.6.5 Reversible Phase Transformation ... 61
3.7 Formulation of Second Law for Irreversible Processes ... 63
3.8 Spontaneous or Natural Processes ... 66
3.9 Entropy Change in Irreversible or Natural Processes ... 66
3.10 Degradation of Energy and Increase of Entropy ... 68
3.11 Entropy Change Associated with Chemical Reaction ... 69
3.12 Entropy Change in Isolated Systems ... 70
3.13 Processing of Materials in Solid-State and Entropy Change ... 71

4 Free Energy, Criteria for Equilibrium and Thermodynamic Relations ... 81
4.1 Free Energy ... 81
4.1.1 Helmholtz Free Energy ... 83
4.1.2 Gibbs Free Energy ... 84
4.2 Expressions for the Combined Statements of First and Second Laws of Thermodynamics ... 84
4.3 Criteria of Equilibrium: Thermodynamic Potentials ... 86
4.3.1 Criteria of Equilibrium Under Constant Volume and Entropy Conditions ... 86
4.3.2 Criteria of Equilibrium Under Constant Pressure and Entropy Conditions ... 88
4.3.3 Criteria of Equilibrium Under Constant Temperature and Volume Conditions ... 89

4.3.4 Criteria of Equilibrium Under Constant Temperature and Pressure Conditions ... 90
4.4 Change in Free Energy and Feasibility of a Reaction ... 92
4.4.1 ΔH and ΔS Values, and Feasibility of a Reaction ... 93
4.5 Standard States ... 94
4.6 Dependence of ΔG on Temperature and Pressure ... 95
4.6.1 Derivation of an Equation for ΔG Versus T Relation ... 95
4.6.2 Derivation of an Equation for ΔG Versus P Relation ... 96
4.7 Gibbs–Helmholtz Equations ... 98
4.8 Maxwell's Relations ... 100
4.8.1 Application of Maxwell's Relations: Few Examples ... 102
4.9 Clausius–Clapeyron Equation ... 105
4.9.1 Application of Clausius–Clapeyron Equation ... 106
4.10 Richard's Rule ... 109
4.11 Trouton's Rule ... 109
4.12 Third Law of Thermodynamics ... 111
4.12.1 Nernst's Heat Theorem ... 112
4.12.2 Latest Statement of Third Law of Thermodynamics ... 113
4.12.3 Concept of Complete Internal Equilibrium ... 113
4.12.4 Verification of Third Law of Thermodynamics ... 114
4.12.5 Consequences of Third Law of Thermodynamics ... 115
4.12.6 Significance of Third Law of Thermodynamics ... 117

5 Fugacity, Activity, Equilibrium Constant and Ellingham Diagram ... 125
5.1 Fugacity ... 125
5.1.1 Physical Significance of Fugacity ... 127
5.1.2 Fugacity of Solids and Liquids ... 128
5.1.3 Variation of Fugacity with Pressure ... 128
5.1.4 Variation of Fugacity with Temperature ... 129
5.2 Thermodynamic Activity ... 130
5.3 Concept of Equilibrium Constant ... 132
5.3.1 Relation Between Equilibrium Constant (K) and Standard Free Energy Change (ΔG^0) ... 134
5.3.2 Temperature Dependence of Equilibrium Constant—Van't Hoff Isochore ... 135
5.3.3 Effect of Pressure on Equilibrium Constant in Gaseous Phase Chemical Reactions ... 137
5.4 Ellingham Diagrams ... 138
5.4.1 Construction of Ellingham Diagram for Oxides ... 139
5.4.2 Equilibrium Partial Pressure of Oxygen and Oxidation–Reduction Phenomena in a Metal–Metal Oxide System ... 141
5.4.3 Relative Stabilities of Metal Oxides ... 144
5.4.4 Phase Transformation ... 145

5.4.5 Metallothermic Reduction ... 147
5.4.6 The Oxides of Carbon and Carbothermic Reduction ... 147
5.4.7 Deoxidation of Steel ... 149
5.5 Ellingham Diagram for Sulphides ... 150
5.5.1 Relative Stabilities of Sulphides ... 150
5.5.2 Dissociation of Sulphides ... 152
5.5.3 Metallothermic and Carbothermic Reductions of Metal Sulphides ... 152
5.6 Salient Features of Ellingham Diagrams—A Summary ... 153
5.7 Limitations of the Ellingham Diagrams ... 154

6 Thermodynamics of Solutions ... 167
6.1 Introduction ... 167
6.2 Ideal and Non-ideal Solutions ... 168
6.3 Component Interaction in Solution and Deviation from Ideality ... 170
6.4 Partial Molar Quantities and Gibbs–Duhem Equation ... 172
6.5 Integral and Partial Molar Quantities of Mixing ... 175
6.6 Procedure to Obtain Partial Molar Quantities of Components in a Solution ... 176
6.7 Thermodynamics of Mixing of Components of Solutions ... 177
6.7.1 Free Energy Change in the Formation of Solution or Free Energy Change of Mixing ... 178
6.7.2 Heat of Formation of a Solution or Heat of Mixing of the Components ... 179
6.7.3 Entropy Change of Mixing or Entropy Change in the Formation of an Ideal Solution ... 180
6.7.4 Volume Change in the Formation of an Ideal Solution ... 181
6.7.5 Graphical Method for Determining Partial Molar Properties ... 182
6.8 Excess Function or Molar Excess Property ... 184
6.9 Regular Solutions ... 185
6.10 Henry's Law ... 186
6.10.1 One Weight Percent Standard State ... 188
6.11 Dilute Multicomponent Solutions—Interaction Between Solutes ... 188

Appendix A ... 201

Appendix B ... 203

Appendix C ... 207

References ... 211

Index ... 213

Symbols

a	Activity
C_V, C_P	Heat capacities at constant volume and constant pressure
F	Helmholtz free energy
f	Fugacity
G	Gibbs free energy
G^o	Gibbs free energy at standard state
ΔG	Change in Gibbs free energy
H	Enthalpy
h	Planck's constant
K	Equilibrium constant
n	Number of gram moles
N	Mole fraction
P	Total pressure in a system
p_i	Partial pressure of component i in a solution
q	Amount of heat exchanged
Q	Partial molar value of an extensive thermodynamic property
Q'	An extensive thermodynamic property for the arbitrary quantity of a solution
$\overline{Q_A}$ and $\overline{Q_B}$	Partial molar properties of components A and B in the solution
Q_A^o and Q_B^o	Molar properties of components A and B in their standard (pure) states
R	Universal gas constant
S	Entropy
T	Temperature
U	Internal energy
V	Volume
W	Work done
α	Darken's α-function
γ	Activity coefficient
γ_i^o	Henry's law constant for solute i in a binary solution
η	Efficiency of an engine

μ_i	Chemical potential of component *i* in a solution
ρ	Density of a material

Chapter 1
Introduction

This is an introductory chapter consisting of eight sections. In this chapter, importance of metallurgical thermodynamics or thermodynamics of materials in the later version has been highlighted. Concepts of basic terms like thermodynamic systems, state and path functions, thermodynamic processes, thermodynamics and energy parameters, extensive and intensive properties, state of a system, state variables, etc. have been discussed. A short discussion has also been made on thermodynamic equilibrium and its various types in this chapter. Reversible and irreversible processes have been discussed in detail in Sect. 1.5. Not only this, limitations of classical thermodynamics have also been included in this chapter.

Knowledge of the above-mentioned basic terms is necessary to understand the different laws of thermodynamics. The details of these laws, enthalpy, entropy, free energy, Ellingham diagram and thermodynamics of solutions have been covered in the upcoming chapters. This chapter has ended with some solved and unsolved problems.

1.1 Introduction

Thermodynamics has a widespread application in all branches of science and engineering. The laws of thermodynamics are general, but their applications are different for different disciplines. Literally, study of dynamism of thermal energy is known as thermodynamics. But, the generalized definition of thermodynamics is that it is a subject dealing with energy and its transformation from one form to another. Strictly speaking, the science of interconversion of energies needs to be named as something like energy dynamics to cover all forms of energies, but the same traditional name thermodynamics continues to be referred to the study of interconversion of energies. Thermodynamics has non-atomic approach, i.e. it does not bother about atomic and

S. K. Sahoo et al., *Fundamentals of Metallurgical Thermodynamics*,
https://doi.org/10.1007/978-981-99-6671-4_1

microscopic details of a substance. It arrives at the final results by taking into consideration only of the initial and final states of the systems. Thermodynamics has been broadly classified into classical, statistical and irreversible. We shall confine ourselves to classical thermodynamics only because it has got an extensive application in metallurgical, materials, ceramics and chemical engineering. Classical thermodynamics was developed in nineteenth and early twentieth centuries by the efforts of engineers, physicists and chemists, and consists of 1st, 2nd and 3rd laws of thermodynamics and related mathematical deductions. Classical thermodynamics deals with systems of finite macroscopic sizes containing millions of molecules and ignores the details of structure of materials. Hence, these laws and equations of classical thermodynamics are independent of structures of materials and cannot be applied to changes in materials of infinite dimensions, such as on a universe scale or microsystems of just few atoms. This is the point where its power and usefulness lies. The zeroth law of thermodynamics is the later addition and not recognized in many books.

1.1.1 Historical Perspective of Classical Thermodynamics

1.1.1.1 General Thermodynamics

In 1840, Joule's experiments confirmed the relation between heat and work and led to the concept of mechanical equivalent of heat. Later on, it was generalized as law of conservation of energy, which states that energy can neither be created nor be destroyed and it can be only transformed from one form to another. These developments constituted the basis for the 1st law of thermodynamics.

Mechanical energy can be completely converted into heat, but the reverse is not true, i.e. heat cannot be completely converted into work. The first major explanation in this area was brought about by S. Carnot in 1824. His work finally formed the basis for the formulation of 2nd law of thermodynamics, employing the concept of reversible and irreversible processes. Later in the middle nineteenth century, entropy was proposed as the thermodynamic parameter by Clausius and J. J. Thomson for the quantitative application of the 2nd law. In 1906, Nernst heat theorem came into existence, and later on, it was generalized by Max Planck as 'the entropy of any homogeneous substance at complete internal equilibrium may be taken as zero at 0 K'. This is known as the 3rd law of thermodynamics, and thus, the formulation of classical thermodynamics got complete. Although there is no direct proof for the validity of these thermodynamic laws, their applications to various changes taking place in the universe yield correct results.

1.1.1.2 Chemical Thermodynamics

In nineteenth century, the major developments witnessed in chemistry were (i) Dalton's atomic theory and reaction stoichiometry; (ii) Avogadro's hypothesis; (iii) the concept of chemical equilibrium; (iv) the heats of reactions; (v) Faraday's law of electrolysis; etc., and they provided the foundation of chemical thermodynamics. In the late nineteenth century, J. W. Gibbs proposed his famous free energy function, known as Gibbs free energy. The inclusion of free energy concepts in thermodynamics firmly established the foundation of chemical thermodynamics and the subject got well established by 1930. In short, chemical thermodynamics means thermodynamics applied to chemical reactions and equilibrium using the concepts of above mentioned three laws, entropy and free energy.

1.1.1.3 Metallurgical Thermodynamics

Broadly speaking, the application of chemical thermodynamics to metallurgical processes in extractive metallurgy, phase transformations, phase equilibria, etc. led to the development and growth of metallurgical thermodynamics or its later generation as thermodynamics of materials. The application of thermodynamics to metallurgical systems requires experimental data on enthalpies and free energies of reactions/processes, activities and compositions of components in solutions, phase equilibria, etc. Processing of metals and ceramics is carried out primarily at high temperatures, and experimental measurements of the above-mentioned thermodynamic properties at high temperatures require the availability of high-temperature materials, techniques, apparatus and instruments. Many of these facilities were not available earlier. Therefore, experimental measurements and assessment of thermodynamic data in high-temperature metallurgy and materials area required several decades and the subject of metallurgical thermodynamics acquired maturity as late as 1960–70. Metallurgical thermodynamics provides information about the followings:

- Changes in energy and entropy accompanying chemical and physical changes
- Driving force behind a chemical reaction (viz. reduction, oxidation, de-oxidation, desulphurization, decomposition, etc.)
- Phase transformations and phase equilibria
- Stability of phases
- Calculation of heat requirements of processes, i.e. heat balance
- Prediction of process feasibility and establishment of equilibrium
- Calculation of equilibrium composition of co-existing phases
- Thermodynamic properties of metallurgical solutions
- Surficial and Interfacial phenomena.

1.1.2 Limitations of Classical Thermodynamics

The followings are the limitations of classical thermodynamics:

(i) It is totally silent about the structure of materials.

(ii) It does not give any idea about the rate of change. The rates of processes / reactions are studied under a separate branch of science known as kinetics, which helps to assess the time required for a particular change to occur.
(iii) The mechanism of a reaction (i.e. whether chemical, diffusion or mixed control) is not predictable by thermodynamics.
(iv) The thermodynamics deals with only conventional forms of energy, e.g. thermal, mechanical, electrical, chemical, etc. The non-conventional forms of energy related to atomic and subatomic particles (e.g. nuclear energy) are dealt with separately because of need of consideration of all matters as energy, as per the famous Einstein equation. $E = mC^2$.

1.1.3 Thermodynamics and Energy Parameters

There must be some energy parameters to assess the amount of energy exchanged during any process. The thermodynamic science evolves several such energy parameters to reveal the energy exchanged in the processes. The following examples can be cited as illustrations:

(i) A system takes heat (q) and does some work (W). The thermodynamics calculates the balance energy (i.e. $q - W$) in terms of change in internal energy which is an energy parameter.
(ii) A liquid is converted into vapour at the boiling point, and latent heat is provided for this change. The question is that in what form the energy of vapourization exists as a result of boiling. The thermodynamic science evaluates this in terms of a property called entropy of a system which is an energy parameter.
(iii) A metal combines with oxygen gas to form a metal oxide. Here, the question is that in what form the energy is lost from the materials. The thermodynamic science assesses it in terms of either enthalpy or free energy which are another energy parameters.
(iv) When electrolysis or an electrochemical reaction takes place, the thermodynamics assesses the exchange of energy in terms of enthalpy and free energy.

1.2 Thermodynamics Systems

Any portion of the material or space selected for consideration is known as the system while the rest is its surrounding. A thermodynamic system must be chemically stable. If the system is continuously undergoing some chemical change, it cannot be considered as a thermodynamic system. For example, living beings are not systems because they are continuously under the process of some changes. In short, there must be a complete internal chemical stability for anything to be called a thermodynamic system. The thermodynamic systems are commonly classified in the following three ways:

On the basis of Interaction of the System with its Surrounding

(i) An Open System: It exchanges both matter and energy with its surrounding, e.g. vapourization or solidification of liquid.
(ii) A Closed System: It exchanges only energy but not the matter with its surrounding, e.g. a gas enclosed in a cylinder fitted with a piston.
(iii) An Isolated System: It neither exchanges energy nor the matter with its surrounding. It is of theoretical importance only because complete isolation is impossible.

Hence, the mass remains fixed for closed and isolated systems during occurrence of a process. Thermodynamic relations are mostly derived by assuming a closed system.

Based on Material Distribution

(i) Homogenous System: A system is said to be homogeneous if it is chemically uniform (i.e. same composition of constituents) throughout and is generally made up of a single phase, e.g. liquid metal or slag.
(ii) Heterogeneous System: A system is said to be heterogeneous if it is not chemically uniform throughout and generally consists of two or more phases, e.g. mixtures of liquid metal and slag, liquid metal and gas.

On the Basis of Compositions

(i) Single-component System: It consists of only one substance (may be an element or a compound), e.g. water and its vapour.
(ii) Multicomponent System: It consists of two or more chemically and/or physically different entities, e.g. liquid steel, liquid slag.

1.3 State of a System

The state of a system at any instant is described by specifying some of its experimentally determinable properties/variables, e.g. temperature, pressure, volume, composition, density, viscosity, surface tension, etc. The minimum number of properties/variables required to describe the state of a system are called independent state variables/properties. Temperature, pressure and volume are the most common state variables. For a single component system, only two variables out of the three (viz. temperature, pressure and volume) are required for its complete descriptions. In case of a multicomponent system, the independent variables are composition and two of the three variables (viz. temperature, pressure and volume). All other properties (e.g. density, viscosity, surface tension) whose values get fixed with the specification of independent state variables are referred to as dependent state variables. There are various interrelations between the state variables, namely temperature (T), pressure (P) and volume (V). These are called equations of state ($PV = RT$, etc.).

Table 1.1 Difference between extensive and intensive properties/variables

Extensive property/variable	Intensive property/variable
It is a function of the quantity of matter present in the system	It is independent of the quantity of matter present in the system
Extensive properties of a system are additive. For example, total volume of a system is the sum of volumes of all its component parts ($V = V_1 + V_2 + V_3 + \cdots$)	These properties of a system are not additive. For example, temperature/pressure of a system is not the sum of the temperatures/pressures of its component parts ($T \neq T_1 + T_2 + T_3 + \cdots$ and $P \neq P_1 + P_2 + P_3 + \cdots$)
Examples are volume, mass, energy, etc	Examples are temperature, pressure, specific heat, density, co-efficient of thermal expansion, thermal conductivity, etc

1.4 Extensive and Intensive Properties

State properties/variables (dependent or independent) are of two types: (*i*) extensive and (*ii*) intensive. They could be well understood by their comparison given in Table 1.1. Product of intensive and extensive properties is an extensive property, whereas the ratio of any two extensive properties yields an intensive property. The values of extensive properties, expressed as per unit mass of the system/substance, fall in the category intensive properties like specific heat, specific volume, molar energy, molar volume, etc.

1.5 Thermodynamic Processes

When a change of state takes place in a system, under externally or internally imposed constraints, the system is said to undergo a process. When a system undergoes the same changes of state, but with different imposed constraints, the processes are different although the paths may be identical. For example, if iron combines with oxygen to form its oxide, it is a simple chemical process, but when the energy exchanged in this process is taken into account then this process of oxidation becomes a thermodynamic process. There are different types of thermodynamic processes depending upon the condition imposed on the system, and they have been named accordingly. Some such important processes are as follows:

Isothermal Process: A process in which the temperature of the system remains constant during a change of state (i.e. $\mathrm{d}T = 0$) is called a constant temperature or isothermal process.

Isochoric Process: A process which proceeds without any change in volume of the system during a change of state (i.e. $\mathrm{d}V = 0$) is called a constant volume or isochoric process.

Isobaric Process: A process which is accomplished under constant pressure during a change of state (i.e. $\mathrm{d}P = 0$) is called a constant pressure or isobaric process.

Table 1.2 Thermodynamic processes and their characteristics

Process	Characteristics	P–V–T relationship	Reversible work done	Reversible heat exchange
Isobaric	$dP = 0$	$\frac{V}{T} = \text{Constant}$	$P\ (V_f - Vi)$	$C_p(T_f - Ti)$
Adiabatic	$\delta q = 0$	$PV^{\gamma} = \text{Constant}$	$\frac{P_f V_f - P_i V_i}{1-\gamma}$	0
Isothermal	$dT = 0$	$PV = \text{Constant}$	$RT \ln \frac{V_f}{V_i}$	$RT \ln \frac{V_f}{V_i}$
Isochoric	$dV = 0$	$\frac{P}{T} = \text{Constant}$	0	$C_v\ (T_f - Ti)$
Polytropic	–	$PV^n = \text{Constant}$	$\frac{P_f V_f - P_i V_i}{1-n}$	$C_v\ (T_f - Ti)$. $\left(\frac{P}{T} = \frac{\gamma - n}{1-n}\right)$

Adiabatic Process: A process which is carried out without any exchange of heat between the system and surrounding during a change of state (i.e. $\delta q = 0$) is called an adiabatic process.

Polytropic Process: A process which obeys the equation PV^n = Constant (where n is any positive number) is called a polytropic process.

The characteristics of these processes have been listed in Table 1.2.

1.5.1 Reversible and Irreversible Processes

The change or transformation in a system may be reversible (under-equilibrium) or irreversible (i.e. non-equilibrium or spontaneous). A process is said to be reversible when its direction can be reversed (by retracing its path) at any point, leading to the restoration of the initial state of the system. A reversible process passes (hypothetically) through a series of equilibrium stages, i.e. equilibrium → non-equilibrium → equilibrium → non-equilibrium → equilibrium, and thus the process continues. A reversible process is very slow and mostly impracticable, but the concept allows us to handle many practical problems. No any going process is reversible in strict sense. The thermodynamics, presently being studied, is for reversible processes only because they correspond to maximum efficiency. Irreversible process is a natural process which involves spontaneous movement of a system from a non-equilibrium to an equilibrium state.

The basic concept of these two processes can be explained in some detail by taking an example of expansion/compression of an ideal gas contained in a cylinder fitted with a piston. In expansion from state 1 to state 2, the system (gas) does work by transferring its mechanical energy to the surrounding and is given by the following expression:

$$W_{1\to 2} = \int\limits_{V_1}^{V_2} P\mathrm{d}V \tag{1.1}$$

Let us now consider the compression of a gas from volume V_2 to V_1 keeping the conditions of the process to be the same as that for expansion. The work done under this condition will be as follows:

$$W_{2\to 1} = \int\limits_{V_1}^{V_2} P\mathrm{d}V = -\int\limits_{V_1}^{V_2} P\mathrm{d}V \tag{1.2}$$

This indicates that same amount of mechanical energy will have to be spent on the system to bring it to the original state (i.e. to reverse the process). Such processes are often called reversible processes. In case of reversible processes, work done in expansion is equal and opposite to that done in compression, provided both the forward and backward processes are performed under similar conditions. In actual practice, it is not possible to obtain complete reversibility. However, one may approach towards it. The reason lies in the fact that some amount of friction is always present in the system and a part of the energy is lost in overcoming this friction. This introduces irreversibility in the process. Hence, friction should be absent to attain complete reversibility.

In case of irreversible expansion, the pressure exerted on the piston is less than the pressure exerted during reversible expansion. Hence, the curve for irreversible expansion lies below that for the reversible expansion. This has been shown schematically in Fig. 1.1 by curve 1. It can be seen in this figure that work done by the system in irreversible expansion is less than that for reversible process ($W_{\text{rev.}} > W_{\text{irrev}}.$). In case of irreversible compression, the pressure exerted by the piston on the gas is more than the pressure during the reversible process. Hence, the curve for irreversible compression lies above that for reversible process, as shown in Fig. 1.1 (curve 3). It can be seen in this figure that the area under curve 3 is more than that under curve 2, indicating relatively more work done in irreversible compression (i.e. $W_{\text{rev.}} < W_{\text{irrev}}$). As is evident from Fig. 1.1, work done in irreversible compression is also more than that done in irreversible expression ($W_{\textit{irrev.compr.}} > W_{\text{irrev.exp}}.$). It may be pointed out here that the difference of work done between reversible and irreversible processes will increase with an increase in velocity of the piston, i.e. rate of the process.

The above discussion on reversible and irreversible processes has been made in reference to an ideal gas, isothermal process and mechanical energy, but these arguments are true for all forms of energy, other substances (i.e. non-ideal gas, liquid, solid, etc.) and other processes also.

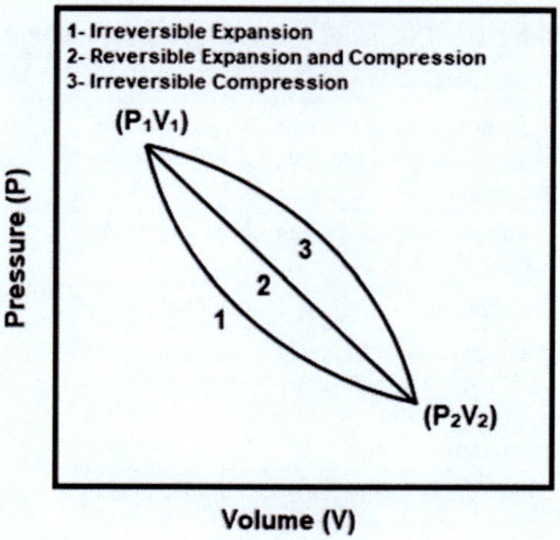

Fig. 1.1 P–V diagram for isothermal reversible and irreversible processes in context to expansion/compression of an ideal gas; path is hypothetical for irreversible cycle

1.6 State and Path Functions

A state function is defined as one whose change in value depends only on the initial and final states of the system, and not on the object's history and path adopted to bring this change. The functions, which not only depend on the initial and final states of the system but also on the path followed by the system in going from one state to other, are called path functions. State functions can be thought of as integrals because integrals depend on only three things: the function, and lower and upper limits. Similarly, state functions as well depend on three things: the property, and initial and final values. On the other hand, multiple integrations within multiple limits are required to take the integrals of path functions.

For a function X to be called a state function, the necessary mathematical condition is $\oint \mathrm{d}X = 0$ whereas for a path function Y, $\oint \delta Y \neq 0$. The symbol $\oint$ denotes the integral over the entire cyclic change. Let us consider the function $Z = f(x, y)$ and further derived expression $\mathrm{d}Z = M\mathrm{d}x + N\mathrm{d}y$, where M and N are the functions of two independent variables of the system x and y. For the function Z to be a state variable, the necessary and sufficient criteria are as follows:

$$\left(\frac{\delta M}{\delta y}\right)_x = \left(\frac{\delta N}{\delta x}\right)_y \tag{1.3}$$

and

$$\left(\frac{\delta Z}{\delta x}\right)_y \left(\frac{\delta x}{\delta y}\right)_z \left(\frac{\delta y}{\delta Z}\right)_x = -1 \tag{1.4}$$

This is a very important condition and is often used in thermodynamics to (*i*) find out whether a particular function of interest is a state property or not and (*ii*) derive a

Table 1.3 Difference between state and path functions

State function	Path function
It is independent of the path followed and depends only on the initial and final states of the system	In addition to the initial and final states of the system, it depends on the path followed also
It can be integrated using initial and final values	It needs multiple integrals and limits for integration
Multiple paths result in same value	Multiple paths result in different value
Examples are enthalpy (H), internal energy (U), Gibb's free energy (G), Helmholtz free energy (F), entropy (S), pressure (P), temperature (T), etc	Examples are mechanical work (w), heat (q), etc
State functions are normally represented by uppercase letters, e.g. H, U, G, F, S	Path functions are normally represented by lowercase letters, e.g. w, q
It describes the equilibrium state of a thermodynamic system, for example, enthalpy (H), internal energy (U), Gibb's free energy (G), Helmholtz free energy (F), entropy (S)	Path functions do not describe the equilibrium state of a thermodynamic system
For an infinitesimal (very small) change in value, state functions are written in their exact differential forms, e.g. $\mathrm{d}U$, $\mathrm{d}H$, $\mathrm{d}S$, $\mathrm{d}G$	For an infinitesimal (very small) change in value, path functions are written in their partial differential forms, e.g. δq, δw
For a finite change in value, state functions are represented by the symbols ΔP, ΔT, ΔV ΔU, ΔH, ΔS, ΔG, etc	For a finite change in value, path functions are written in their partial differential forms, e.g. δq, δw
For a function X to be a state quantity, the necessary condition is $\oint \mathrm{d}X = 0$	For any function Y which is not a function of state $\oint \delta Y \neq 0$
In the expression of type $\mathrm{d}z = M\mathrm{d}x + N\mathrm{d}y$, the function z will be a state variable if $\delta M/\delta y = \delta N/\delta x$	Reverse is true for path function

number of thermodynamic relationships. For more details, state and path functions have been differentiated in Table 1.3.

Applications: State functions are commonly encountered in thermodynamics as many of the derived equations involved their use, such as ΔU, ΔH, ΔS, ΔG, ΔF. In addition, state functions are crucial in thermodynamics because they make the calculations simple and easy, and allow us to calculate the data that could only be obtained through experiments.

1.7 Thermodynamic Equilibrium

The concept of thermodynamic equilibrium is very important in understanding the usefulness of thermodynamic science. The system tends to achieve a state which does not change with time under a given set of external conditions or constraints.

This most stable state is generally referred to as equilibrium state of the thermodynamic system. In a state of thermodynamic equilibrium, flow of matter / energy and phase changes does not occur within the system. The earliest studies on equilibrium were confined only to the thermal state of the system and led to the formulation of zeroth law of thermodynamics. The original zeroth law of thermodynamics states that when two objects are separately in thermal equilibrium (i.e. at the same temperature) with a third object, they are in thermal equilibrium with each other also. But, complete thermodynamic equilibrium means equilibrium with respect to all potentials like thermal, mechanical, chemical, etc. Hence, the generalized statement of the above law for an overall equilibrium is that when two objects are separately in thermodynamic equilibrium with a third object with respect to any one or more of the thermodynamic potentials, they are in equilibrium with each other also. The various types of equilibrium are as follows:

Thermal Equilibrium: Thermal equilibrium means temperature remains uniform throughout the system or two systems are said to be in thermal equilibrium when their temperatures are same and there is no flow of heat between them. Insulating materials (in contact with each other) reach the state of equilibrium after a very long period of time while the conducting materials come in thermal equilibrium very quickly.

Mechanical Equilibrium: Mechanical equilibrium means pressure equilibrium, i.e. uniformity of pressure throughout the system or two systems are in mechanical equilibrium when their pressures are the same.

Chemical Equilibrium: A system is said to be in chemical equilibrium when the rates of all the chemical reactions taking place in the system at any instant are exactly equal in forward and backward directions. This means that the concentrations of reactants and products do not change with passage of time under the state of chemical equilibrium. This definition of chemical equilibrium is based on the considerations of kinetic aspects (i.e. reaction rates) of the reactions. The concept of chemical equilibrium can also be explained through thermodynamic considerations. There are some thermodynamic potential (such as G, F, S and H) which become the necessary and sufficient criteria for the establishment of equilibrium within a system under specified conditions. For example, Gibbs free energy is the necessary and sufficient criteria (i.e. $\Delta G_{T,P} = 0$) for the attainment of equilibrium in a system under thermal and mechanical conditions (i.e. at constant T and P).

It is not necessary that all aspects of thermodynamic equilibrium are reached simultaneously, i.e. some can be established before others. For purely physical processes, such as expansion/compression of a substance, thermodynamic equilibrium consists of mechanical and thermal equilibria. But for physico-chemical processes, thermodynamic equilibrium will also require the attainment of chemical equilibrium in addition to mechanical and thermal equilibria. The equilibrium states of a system can be classified as (*i*) stable state, (*ii*) metastable state and (*iii*) neutral state. These can be illustrated as follows:

Stable Equilibrium: A system in stable equilibrium gets displaced from its equilibrium state on application of an external constraint, but returns back to its original state on the removal of an applied constraint.

Metastable Equilibrium: If displacement under external constraint makes the system stable in some other state then the system is said to be in metastable equilibrium. In metastable state condition, the system is at partial or pseudo-equilibrium and the system undergoes a change so slowly that there is no perceptible change in a long period of time (say, several years/centuries/millennia). Cementite (Fe_3C) in iron–carbon system is metastable, but it can stay for perhaps several centuries or millennia at room temperature.

Neutral Equilibrium: On displacement, if the system remains in displaced state then it is said to be in neutral equilibrium.

1.8 Concluding Remarks

The main conclusions drawn from this chapter may be summarized as follows:

(i) The present text is primarily concerned with classical thermodynamics because majority of the applications in metallurgical, materials, ceramics and chemical engineering are based on it. The term thermodynamics, in general, refers to classical thermodynamics.

(ii) The three laws of thermodynamics and concepts of auxiliary functions (e.g. enthalpy, entropy, free energy, activity, fugacity, equilibrium constant) and their relations constitute the foundations of thermodynamics.

(iii) Thermodynamics deals with conservation of energy and its transformation from one form to another.

(iv) The application of thermodynamics contributes greatly in understanding the scientific principles behind the energy requirements in industrial manufacturing activities. The energy resources are limited, and this understanding helps in reducing the energy consumption.

(v) Metallurgical thermodynamics or its later generation thermodynamics of materials belongs to chemical thermodynamics and provides information on process feasibility under prescribed conditions, heat balance for a process, phase transformations, stability of phases, establishment of equilibrium, thermodynamic properties of metals and alloys, interfacial phenomena, etc.

(vi) The concepts of the basic terms like system, state and path functions, extensive and intensive properties, thermodynamic processes, etc. are required to understand the thermodynamic laws and have been covered in this chapter.

(vii) Any selected portion of this infinite universe having real or imaginary dimensions is known as a thermodynamic system, and the rest of the universe is its surrounding. A thermodynamic system can be classified on the basis of its (i) interaction with the surroundings, (ii) own material distribution, and (iii) own composition.

(viii) The extensive state properties or variables of a system vary with its mass or size, whereas the intensive state properties of a system are independent of its amount or size.

(ix) For a system to be in the state of complete equilibrium (i.e. thermal, mechanical and chemical), the temperature, pressure and chemical potential of each of the components must be same throughout the system.

Solved Problems

Problem 1.1 Two moles of an ideal gas at 773 K expands isothermally from a pressure of 11.5 to 1.5 atm. under reversible condition. Find out the work done in this expansion process.

Solution

We know that
$$W = nRT \ln \frac{V_2}{V_1}$$
$$= 2 \times 1.98 \times 773 \ \ln (11.5/1.5)$$
$$= 6235\ cal = 26087\ J$$

Problem 1.2 A reversible polytropic process follows the equation $PV^{\eta} =$ constant. Prove that the work done (W) in this process for a change of pressure and volume from P_1 to P_2 and V_1 to V_2 is given by the following expression:

$$W = \frac{P_2V_2 - P_1V_1}{1 - \eta}$$

Solution

As per question, $PV^{\eta} =$ constant (say I). We know that $\delta W = P\mathrm{d}V$.

or $W = \int_{V_1}^{V_2} \frac{I}{V^{\eta}} \mathrm{d}V = I\left[\frac{V^{1-\eta}}{1-\eta}\right]_{V_1}^{V_2} = \frac{IV_2^{1-\eta} - IV_1^{1-\eta}}{1-\eta}$.

Since, $P_1V_1^{\eta} = P_2V_2^{\eta} = I$. Hence, the above equation for W becomes as follows:

$$W = \frac{P_2V_2^{\eta}V_2^{1-\eta} - P_1V_1^{\eta}V_1^{1-\eta}}{1 - \eta} = \frac{P_2V_2 - P_1V_1}{1 - \eta} \text{ (proved)}$$

Problem 1.3 One mole of an ideal gas at a temperature of 673 K expands from a pressure of 20 atm to 2 atm and the final temperature of the gas becomes 373 K. Find out the value of work done in this process by considering the transfer from initial to final state to be reversible.

Solution

For an ideal gas $P_1V_1 = nRT_1$ or Initial volume of the gas $(V_1) = \frac{nRT_1}{P_1}$.

or $V_1 = \frac{1 \times 0.082 \times 673}{20} = 2.76\,\mathrm{L}$

Similarly, final volume of the gas $(V_2) = \frac{nRT_2}{P_2} = \frac{1 \times 0.082 \times 373}{2} = 15.29\,\mathrm{L}$.

Let us assume that the expression, $PV^{\eta} =$ constant, represents the reversible process, then

$$P_1 V_1^\eta = P_2 V_2^\eta \quad \text{or} \quad (V_1/V_2)^\eta = P_2/P_1$$

or $\eta \ln(V_1/V_2) = \ln(P_2/P_1)$

or $\eta \ln(P_2/P_1)/\ln(V_1/V_2) = \ln(2/20)\ln(2.76/15.29) = 1.35$

From the equation derived in problem 1.2, we know that $W = \dfrac{P_2V_2 - P_1V_1}{1-\eta} = \dfrac{2 \times 15.29 - 20\times}{1 - 1.35}$

$= 70.34$ L.atm/Kmol $= 7128$ J/Kmol

[Because $R = 0.082$ L.atm/Kmol $= 8.314$ J/Kmol]

Problem 1.4 1.5 mol of an ideal gas at a temperature of 973 K expands under adiabatic condition and its volume increases from 4.3 to 11.3 L. Find out the value of work done in this process by assuming the expansion to be reversible. Take $\eta = 1.4$.

Solution

As per question, $V_1 = 4.3$ L and $V_2 = 11.3$ L.

For an ideal gas $P_1V_1 = nRT$ or $P_1 = \frac{nRT}{V_1} = \frac{1.5\times0.082\times973}{4.3} = 27.83$ atm

For an adiabatic process, $P_1V_1^\eta = P_2V_2^\eta$

or $P_2 = \frac{P_1V_1^\eta}{V_2^\eta} = \frac{P_1V_1^\eta\left(\frac{V_1}{V_1}\right)}{V_2^\eta} = \frac{P_1V_1V_1^{\eta-1}}{V_2^\eta} = \frac{nRTV_1^{\eta-1}}{V_2^\eta} = \frac{1.5\times0.082\times973\times(4.3)^{1.4-1}}{(11.3)^{1.4}}$

or $P_2 = 7.2$ atm

From the equation derived in Problem 1.2, we know that $W = \dfrac{P_2V_2 - P_1V_1}{1-\eta} = \dfrac{7.2 \times 11.3 - 27.83 \times 4.3}{1 - 1.4}$

$= 95.77$ L.atm/Kmol $= 9710$ J/Kmol

Exercise

1.1 Two moles of an ideal gas expands reversibly at a temperature of 473 K and its pressure drops from 10 atm to 4.55 atm at constant temperature followed by further expansion to 1.5 atm under adiabatic condition. Find out the value of total work done in this process. Take $\eta = 1.4$.

[**Ans**: 9443 J]

Chapter 2
First Law of Thermodynamics

Entire discussion in this chapter has been covered in six sections. In this chapter, concepts of internal energy (U), enthalpy (H), amount of heat exchanged between system and the surrounding (Δq), work done on or by the system (δw), heat capacities of the substance at constant pressure and volume (C_P and C_V), interconversion of energy, etc. have been elaborated substantially. A detailed knowledge on statements of first law of thermodynamics and its application for various thermodynamic processes, e.g. isothermal, isochoric, isobaric and adiabatic, have been provided in Sect. 2.2. Mathematical equations of this first law for these processes were derived and discussed. Equations for (i) relation between C_P and C_V and (ii) variation of heat capacity and enthalpy change (ΔH) with temperature have been established. Not only these, some other correlations between the above-mentioned thermodynamic properties have also been established.

Enthalpy changes associated with chemical reactions, phase transformations with and without change in state of the system and mixing of components have been described sufficiently in Sect. 2.4. This chapter also deals with a short discussion on standard state, and Hess's and Kirchhoff's laws. At the end of this chapter, many important problems have been solved and some given as exercise.

2.1 Concept of Internal Energy

The energy stored in the system by virtue of the configuration and motion of the molecules constituting the system is called its internal energy. It is a state property of the system and is denoted by the symbol U. The energy due to mass motion of the system as a whole (i.e. kinetic energy) and that due to its external position in a gravitational, electrical or magnetic field (i.e. potential energy) are not included in the internal energy as these forms of energy are not the properties of the system. The molecules constituting the system possess kinetic energy of translation, rotation and

S. K. Sahoo et al., *Fundamentals of Metallurgical Thermodynamics*,
https://doi.org/10.1007/978-981-99-6671-4_2

vibration. They also possess potential energy due to the forces of attraction among them. These molecular potential and kinetic energies contribute to the internal energy of the system. It may also be noted that the internal energy of a material depends on its mass, chemical compositions and structure, and temperature. For a material of fixed mass, composition and structure, it is a function of temperature only. The addition of heat to the system results in an increase of its molecular kinetic energy and thus internal energy also.

Let, the quantity of heat supplied to the system is q and as a consequence, the system does w amount of work on the surrounding. Law of conservation of energy demands that q must be equal to w. But, experiments have revealed that for non-cyclic processes, $q \neq w$ or $q - w \neq 0$. In order to save the law of conservation of energy, it was proposed that a system possesses an internal energy U, which is a hidden form of energy stored in it. The difference between q and w was attributed to change in internal energy (i.e. ΔU) during the process, i.e.

$$\Delta U = q - w \tag{2.1}$$

For non-cyclic processes (i.e. when $q \neq w$), we have two distinct possibilities as follows:

Case I: When $q < w$, some amount of already stored energy gets removed from the system during a non-cyclic process. In other words, the system partly imparts some amount of its energy to do the work.

Case II: When $q > w$, a part of the total supplied energy gets absorbed in the system and the remaining is utilized in doing work on the surrounding. These changes in the stored energy (by an amount $q - w$) are nothing but the change in internal energy of the system.

It also needs to be understood that it is impossible to measure the absolute value of internal energy of a substance. We can determine only change in internal energy (i.e. ΔU). This is not a serious limitation because changes in internal energy are usually required in thermodynamic analysis. It has been proved experimentally that ΔU (i.e. $q - w$) is zero for a cyclic process. Thermodynamic observations have shown that the value of $q - w$ remains constant as long as the initial and final states are not changed. In other words, the value of the difference $q - w$ depends on the initial and final states of system and not on the path followed, although both of them (q and w) individually are path functions. All these experimental findings lead to the conclusion that the function U, i.e. $(q - w)$, is a state function. This conclusion holds good whatever the process is (reversible or irreversible). Like potential energy, the internal energy can be stored in the system. But, heat and work cannot be stored or conserved independently because they depend on the process.

2.1.1 *Internal Energy in Terms of Partial Derivatives*

In thermodynamics, we are generally concerned with *P*, *V* and *T* as state variables. In chemical thermodynamics, the chemical composition and structure of substance in the system are also variables. For a substance of fixed chemical composition and structure throughout the process, *P*, *V* and *T* become the variables. Out of *P*, *V* and *T*, only two are independent variables and the third gets automatically fixed. Hence, for a substance/system of fixed chemical composition and structure,

$$U = f(V, T) \tag{2.2}$$

or

$$U = f(P, T) \tag{2.3}$$

or

$$U = f(P, V) \tag{2.4}$$

From the fundamental theorem of partial differentiation, the change in internal energy can be expressed as follows:

$$\mathrm{d}U = \left(\frac{\delta U}{\delta V}\right)_T \mathrm{d}V + \left(\frac{\delta U}{\delta T}\right)_V \mathrm{d}T \tag{2.5}$$

or

$$\mathrm{d}U = \left(\frac{\delta U}{\delta P}\right)_T \mathrm{d}P + \left(\frac{\delta U}{\delta T}\right)_P \mathrm{d}T \tag{2.6}$$

or

$$\mathrm{d}U = \left(\frac{\delta U}{\delta P}\right)_V \mathrm{d}P + \left(\frac{\delta U}{\delta V}\right)_P \mathrm{d}V \tag{2.7}$$

Similarly, for a reaction undergoing some change in chemical composition (without any change in structure) at constant pressure, volume, temperature and composition (*N*) become the variables. Therefore, *U* can be expressed as follows:

$$U = f(V, T) \tag{2.8}$$

or

$$U = f(N, T) \tag{2.9}$$

or

$$U = f(V, N) \tag{2.10}$$

Like Eqs. (2.5) to (2.7), dU can be expressed as follows:

$$\mathrm{d}U = \left(\frac{\delta U}{\delta V}\right)_T \mathrm{d}V + \left(\frac{\delta U}{\delta T}\right)_V \mathrm{d}T \tag{2.11}$$

or

$$\mathrm{d}U = \left(\frac{\delta U}{\delta N}\right)_T \mathrm{d}N + \left(\frac{\delta U}{\delta T}\right)_N \mathrm{d}T \tag{2.12}$$

or

$$\mathrm{d}U = \left(\frac{\delta U}{\delta V}\right)_N \mathrm{d}V + \left(\frac{\delta U}{\delta N}\right)_V \mathrm{d}N \tag{2.13}$$

Each of the bracketed differential terms expresses the rate of change of U with respect to the variable in the denominator when other variables are held constant. Any differential property which can be expressed in this way is called a complete or an exact differential property. These equations in which internal energy has been expressed as a function of independent state variables are called "Caloric equations" of state, and they are helpful in evaluation of heat capacities.

2.2 Statements of the First Law of Thermodynamics

The first law of thermodynamics is mainly based on the law of conservation of energy and also deals with its transformation from one form to another. The statements can be enunciated as follows:

(i) *Conservation of Energy*: Energy can neither be created nor be destroyed. Total energy of the system and its surroundings remains constant during a thermodynamic process (i.e. $E_{\mathrm{System}} + E_{\mathrm{Surrounding}} = \mathrm{Constant}$). In other words, during an interaction between the system and its surroundings, the amount of energy gained by the system is exactly equal to the amount of energy lost by the surroundings.

(ii) *Transformation of Energy*: Energy can be transformed from one form to another.

All the above statements follow Eq. (2.1), which is generally written as $\Delta U = q - w$ for a finite change in the process. For an infinitesimal (i.e. very small) change in state, the first law can be mathematically expressed as follows:

$$\mathrm{d}U = \delta q - \delta w \tag{2.14}$$

where dU—change in internal energy of the system; δq—amount of heat absorbed by the system from the surrounding; δw—work done by the system.

Infinitesimal changes in q and w are designated by δq and δw not by dq and dw because they are path functions. For mathematical convenience, it is preferable that the Eq. (2.1) be expressed in terms of differential quantities, i.e. Equation (2.14). Out of δq amount of heat energy supplied to the system, energy equivalent to δw be spent in doing the work δw. Hence, $(\delta q - \delta w)$ amount of energy is only responsible for change in internal energy of the system. Equation (2.14) can be further written as:

$$dU = \delta q - Pdv - \delta w' \tag{2.15}$$

where $\delta w = Pdv + \delta w'$; Pdv is the mechanical work and $\delta w'$ represents other forms of work like electrical, magnetic, gravitational, etc. The normal chemical and metallurgical systems do only mechanical work. Therefore, $\delta w'$ is assumed to be zero and the above equation becomes $dU = \delta q - \delta w$.

2.2.1 Statements of the First Law of Thermodynamics for Different Systems

One can arrive at the following statements of this law for isolated, open and closed systems. For an isolated system which neither exchanges energy nor mass with its surroundings, the first law takes the form as: the total energy of an isolated system remains constant, though it may change its form partly or completely. Closed and open systems are capable of exchanging energy with their surroundings. For these systems, one can state this law in the form "the total energy of the system and its surroundings remains constant during a thermodynamic process" though it can change its form and can get transferred from system to the surrounding.

2.2.2 Implications of the First Law of Thermodynamics

In this section, some applications of the first law of thermodynamics in different processes have been outlined. In all these thermodynamic processes, working substance of the system has been considered to be an ideal gas for which

$$\text{PV} = \text{RT (for one mole)} \tag{2.16}$$

and

$$\left(\frac{\delta U}{\delta V}\right)_T = 0 \tag{2.17}$$

As evident from Eq. (2.17), the internal energy of an ideal gas does not depend on volume at fixed temperature. The discussions for isothermal, isochoric, isobaric and adiabatic processes are as follows:

(i) *Isothermal Processes*: During the course of these processes, temperature of the system does not change. As evident from Eq. (2.17), the internal energy of an ideal gas is a function of temperature only. As temperature increases, U increases and vice versa. Hence, under isothermal condition, $\mathrm{d}U = 0$, i.e. an internal energy of the system remains unchanged. In such a condition, the first law of thermodynamics becomes $\delta q = \delta w$ (i.e. $P\mathrm{d}V$).

Work done or heat exchanged during an isothermal process for a volume change in the system (containing one mole of an ideal gas) from V_i to V_f can be expressed by the following expression:

$$\delta w = \delta q = \mathrm{RT}\ \ln \frac{V_f}{V_i} \tag{2.18}$$

This equation shows that for $V_f > V_i$ (i.e. for expansion of gas), δq is positive (endothermic), i.e. the system will absorb heat from its surroundings and produce an equivalent amount of work. On the other hand, for $V_f < V_i$ (i.e. for compression of gas), δq is negative (exothermic), i.e. the system will release heat into its surroundings when mechanical work is done on it.

One knows that the enthalpy,

$$H = U + \mathrm{PV} \tag{2.19}$$

When the pressure of the system changes from P_i to P_f and the corresponding change in volume is from V_i to V_f, the equation for enthalpy change can be written as follows:

$$\Delta H = \Delta U + P\Delta V = \Delta U + \left(P_f V_f - P_i V_i\right) \tag{2.20}$$

For one mole of an ideal gas under isothermal condition, $P_f V_f = P_i V_i = \mathrm{RT}$.

Hence, Eq. (2.20) yields $\Delta H = 0$. The net conclusion is that all the processes, operating under isothermal condition, are accompanied without any change in internal energy and enthalpy (i.e. $\Delta U = 0$ and $\Delta H = 0$), and work done by the system is equivalent to the heat input.

(ii) *Isochoric Processes*: They are the processes in which the volume remains unchanged (i.e. $\mathrm{d}V = 0$). In such processes, the work done ($P\mathrm{d}V$) is zero and the first law of thermodynamics becomes $\mathrm{d}U = \delta q$. This indicates that under isochoric condition, the heat energy supplied to the system is utilized in increasing its internal energy. It is also evident from this equation that heat exchanged behaves like a state function under constant volume condition.

(iii) *Isobaric Processes*: Isobaric processes are carried out under constant pressure condition (i.e. $\mathrm{d}P = 0$). Under constant pressure condition, the first law of thermodynamics takes the following form:

$$\mathrm{d}U = \delta q - P\mathrm{d}V$$

or

$$\delta q = \mathrm{d}U + P\mathrm{d}V = \mathrm{d}(U + PV) = \mathrm{d}H$$

or

$$\delta q = \mathrm{d}H \tag{2.21}$$

By looking at whether a process is exothermic or endothermic, one can examine the relationship between δq and $\mathrm{d}H$ in the following manner:

Case I—If the process/reaction is endothermic (i.e. $\delta q > 0$ or + ve), then ΔH is also + ve. This indicates that under constant pressure condition, endothermic reactions are accompanied by an increase of enthalpy.

Case II—If the reaction releases heat to its surroundings (i.e. exothermic and $\delta q < 0$ or – ve) then ΔH is – ve, indicating that exothermic reactions at constant pressure proceed by the decrease of enthalpy.

(iv) *Adiabatic Processes*: Processes that occur without any exchange of heat energy between the system and its surroundings (i.e. $\delta q = 0$) are called adiabatic processes. In such processes, the first law of thermodynamics gets reduced to the followings:

$$\mathrm{d}U = -\delta w = -P\mathrm{d}V \tag{2.22}$$

or

$$\delta w = -\mathrm{d}U \tag{2.23}$$

Equation (2.22) clarifies that the adiabatic work done on the system leads to an increase of its internal energy and as a result the temperature of the system increases. Equation (2.23) explains that under adiabatic condition, the system does work at the expense of its own internal energy and therefore, temperature of the system falls.

Since $\mathrm{d}U = C_V\mathrm{d}T$, where C_V is the heat capacity of the system at constant volume. Hence, an adiabatic work done $(P\mathrm{d}V) = -\mathrm{d}U = -C_V\mathrm{d}T$.

For a change of temperature from T_i to T_f, adiabatic work done $= -C_V(T_f - T_i)$.

2.3 Heat Capacity

It is usually defined as the amount of heat required to raise the temperature of a substance, under consideration, by 1 °C. The heat capacity per mole is called the molar heat capacity, and the heat capacity per unit mass of the material is known as specific heat capacity or simply specific heat. Heat capacity depends on the quantity of the substance and is thus an extensive property. When the heat capacity is divided by the amount of substance (i.e. mass or mole), it becomes an intensive property. In thermodynamics, the molar heat capacity (C) is most widely used.

Let δq amount of heat is required to raise the temperature of one mole of the substance by $\mathrm{d}T$. Then,

$$C = \frac{\delta q}{\mathrm{d}T} \tag{2.24}$$

q is a path function, hence the heat capacity is dependent on the path followed by the reaction unless the conditions are specified to define the path. For example, under isochoric condition, $\delta q = \mathrm{d}U$. Since, the internal energy is a state function, so too is the isochoric heat capacity. The molar heat capacity at constant volume (under isochoric condition) is given by the following equation:

$$C_V = \left(\frac{\delta q}{\delta T}\right)_V \tag{2.25}$$

As outlined in Sect. 2.2.2, $\delta q = \mathrm{d}U$ for constant volume processes. Hence, Eq. (2.25) becomes

$$C_V = \left(\frac{\delta q}{\delta T}\right)_V = \left(\frac{\delta U}{\delta T}\right)_V \tag{2.26}$$

C_V is equal to the rate of change of internal energy with temperature at constant volume. From Eq. (2.25), we have the following:

$$\mathrm{d}U = C_V \mathrm{d}T \tag{2.27}$$

On integrating Eq. (2.27) between temperatures T_1 and T_2, we get

$$\Delta U = \int_{T_1}^{T_2} C_V \mathrm{d}T \tag{2.28}$$

This equation relates the change in internal energy with heat capacity at constant volume. Similarly, molar heat capacity at constant pressure (C_P) is given by the following equation:

$$C_P = \left(\frac{\delta q}{\delta T}\right)_P \tag{2.29}$$

As outlined in Sect. 2.2.2, $\delta q = \mathrm{d}H$ for constant pressure processes. Hence, Eq. (2.29) becomes

$$C_P = \left(\frac{\delta q}{\delta T}\right)_P = \left(\frac{\delta H}{\delta T}\right)_P \tag{2.30}$$

C_P is equal to the rate of change of enthalpy with temperature at constant pressure. From Eq. (2.30), we have the following:

$$\mathrm{d}H = C_P \mathrm{d}T \tag{2.31}$$

Let the substance is heated from a temperature of T_1 to T_2 at constant pressure without any change of state or phase transformation. Now, on integration of Eq. (2.31) between temperature limits T_1 and T_2, we get

$$\int\limits_{H_1}^{H_2} \mathrm{d}H = \int\limits_{T_1}^{T_2} C_P \mathrm{d}T$$

or

$$H_2 - H_1 = \int\limits_{T_1}^{T_2} C_P \mathrm{d}T \tag{2.32}$$

This equation relates the change in enthalpy with heat capacity at constant pressure. It may be mentioned here that these equations of heat capacities apply only two homogeneous substances and not to heterogeneous substances/systems, as in the latter case, there may be some other effects due to change in composition of phases with temperature. The value of C_P is always greater than C_V (i.e. $C_P > C_V$) because C_P includes the heat required to do the work also besides raising the temperature. At constant volume condition, heat is required only for raising the temperature and not for doing the work C_V (i.e. $C_P > C_V$).

2.3.1 Relation Between C_P and C_V

We know that $C_P = \left(\frac{\delta q}{\delta T}\right)_P = \left(\frac{\delta H}{\delta T}\right)_P$ and $C_V = \left(\frac{\delta q}{\delta T}\right)_V = \left(\frac{\delta U}{\delta T}\right)_V$.

Therefore, $C_P - C_V = \left(\frac{\delta H}{\delta T}\right)_P - \left(\frac{\delta U}{\delta T}\right)_V = \left[\frac{\delta(U+PV)}{\delta T}\right]_P - \left(\frac{\delta U}{\delta T}\right)_V$

$$= \left(\frac{\delta U}{\delta T}\right)_P + P\left(\frac{\delta V}{\delta T}\right)_P - \left(\frac{\delta U}{\delta T}\right)_V \tag{2.33}$$

From Eq. (2.11), we have the following relationship:

$$\mathrm{d}U = \left(\frac{\delta U}{\delta V}\right)_T \mathrm{d}V + \left(\frac{\delta U}{\delta T}\right)_V \mathrm{d}T$$

On dividing both sides of this equation by $\mathrm{d}T$ at constant pressure, we get the following:

$$\left(\frac{\delta U}{\delta T}\right)_P = \left(\frac{\delta U}{\delta V}\right)_T \left(\frac{\delta V}{\delta T}\right)_P + \left(\frac{\delta U}{\delta T}\right)_V \tag{2.34}$$

On combining Eqs. (2.33) and (2.34), we get the following:

$$\begin{aligned} C_P - C_V &= \left(\frac{\delta U}{\delta V}\right)_T \left(\frac{\delta V}{\delta T}\right)_P + \left(\frac{\delta U}{\delta T}\right)_V + P\left(\frac{\delta V}{\delta T}\right)_P - \left(\frac{\delta U}{\delta T}\right)_V \\ &= P\left(\frac{\delta V}{\delta T}\right)_P + \left(\frac{\delta U}{\delta V}\right)_T \left(\frac{\delta V}{\delta T}\right)_P \\ &= \left(\frac{\delta V}{\delta T}\right)_P \left[P + \left(\frac{\delta U}{\delta V}\right)_T\right] \end{aligned} \tag{2.35}$$

For one mole of an ideal gas, $PV = RT$
Therefore,

$$\left(\frac{V}{T}\right) = \left(\frac{R}{P}\right) \tag{2.36}$$

On differentiating this equation at constant pressure, we have

$$\left(\frac{\delta V}{\delta T}\right)_P = \frac{R}{P} \tag{2.37}$$

Again, internal energy of an ideal gas is a function of temperature only, and it does not change with volume at constant temperature, i.e.

$$\left(\frac{\delta U}{\delta V}\right)_T = 0 \tag{2.38}$$

On combining Eqs. (2.35) and (2.38), we get the following:

$$C_P - C_V = \frac{R}{P} \times P = R$$

Therefore, for an ideal gas,

$$C_P - C_V = R \tag{2.39}$$

2.3.2 Dependence of Heat Capacity on Temperature

Specific/molar heat capacity of the material changes with temperature. As outlined in Eqs. (2.19), (2.26) and (2.30), we have the following relationships:

$$H = U + PV; C_V = \left(\frac{\delta U}{\delta T}\right)_V; C_P = \left(\frac{\delta H}{\delta T}\right)_P$$

For solids and liquids, the *PV* term is very small and can be neglected. Hence, $H \approx U$ and C_P also becomes equal to C_V. In other words, there is no distinction between C_P and C_V so far as applications are concerned. Variation in specific/molar heat capacity of any substance with temperature is complex. However, after a lot of experimental work, empirical relationships for a number of elements and compounds have been developed in the following polynomial form:

$$C_P = a + bT + cT^{-2} \tag{2.40}$$

$$C_V = a' + b'T + c'T^{-2} \tag{2.41}$$

where a, b, c, a', b' and c' are constants and their values depend on the characteristics of the materials under study. These equations are purely empirical, and their use is strictly limited to the range of temperature for which the data of the above constants for a particular substance was derived. The last terms (involving T^{-2}) in Eqs. (2.40) and (2.14) are the smallest and, therefore, often ignored. In some cases (e.g. liquid metals) where data is not sufficient, both the last two terms (involving T and T^{-2}) are usually ignored and C_P or C_V is assumed to be constant. These equations are frequently used for solid metals and compounds at elevated temperatures. However, they may be applied for diatomic and polyatomic gases also. In order to simplify the calculations involving specific/molar heat capacity, a mean value of C_P or C_V over certain temperature range may be used. This mean value is defined as follows:

$$C_P/C_V = \frac{\int_{T_1}^{T_2} (C_P/C_V)\mathrm{d}T}{(T_2 - T_1)} \tag{2.42}$$

where T_1 and T_2 are the extreme temperatures of the system undergoing a process.

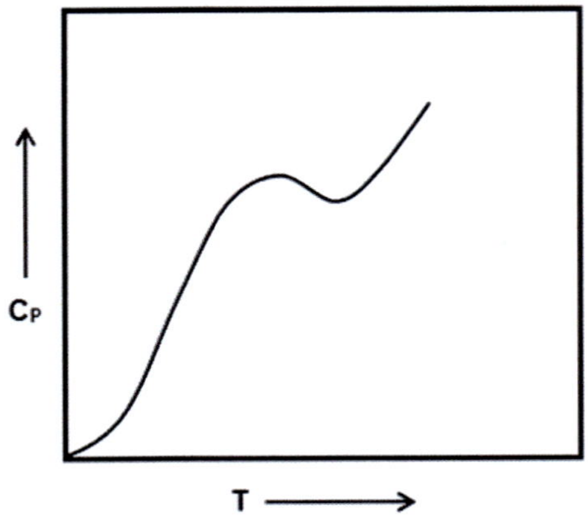

Fig. 2.1 Constant pressure molar heat capacity (C_P) versus temperature plot for nickel

In addition, some empirical rules have also been suggested for approximate calculation of molar heat capacities of substances. For monatomic gases at high temperatures, the molar heat capacity C_P approaches a value of $(5/2)R$, whereas the same is close to $(7/2)R$ for diatomic gases. This generally holds good at room temperature. However, these values of C_P increase with rise of temperature. For solid materials the molar heat capacity C_P is nearly equal to 3*R*.

Figure 2.1 shows the variation of C_P with temperature for nickel in the temperature range of 0 to 800 K (i.e. − 273 to 527 °C). The plot shows a gradual increase in the value of C_P with rise of temperature except for a sharp discontinuity at 600 K (i.e. 327 °C). This discontinuity is due to the magnetic transformation in nickel at 600 K. In other words, discontinuities in such plots occur wherever there is phase, magnetic and order–disorder transformations. Similar plots have been reported for other elements or compounds.

2.3.3 *Importance of Heat Capacities* C_P *and* C_V

C_P is more important and mostly used in metallurgical thermodynamics than C_V because most of the chemical reactions in metallurgical processes occur at constant pressure and not at constant volume. The value of C_P is always greater than that of C_V.

Since,

$$C_V = \left(\frac{\delta U}{\delta T}\right)_V \text{ and } C_P = \left(\frac{\delta H}{\delta T}\right)_P = \left[\frac{\delta(U + PV)}{\delta T}\right]_P = \left(\frac{\delta U}{\delta T}\right)_P + P\left(\frac{\delta V}{\delta T}\right)_P$$

On considering $\left(\frac{\delta U}{\delta T}\right)_V = \left(\frac{\delta U}{\delta T}\right)_P$, C_P is still having an extra term of $P\left(\frac{\delta V}{\delta T}\right)_P$.

Hence, $C_P > C_V \cdot C_P$ values of different substances are very frequently used in calculating enthalpy and entropy changes (i.e. ΔH and ΔS) over a range of temperature. In majority of the chemical and metallurgical processes, the materials are heated and cooled. Knowledge of molar/specific heat capacities helps in calculation of heat required in heating the material over a certain temperature range.

2.4 Enthalpy

When the system undergoes a change at constant pressure, the quantity $U + PV$ occurs very frequently and is called enthalpy of the system (H), i.e. $H = U + PV$. In other words, enthalpy (H) is the sum of internal energy (U) and capacity of the system to do work (PV). Enthalpy can also be expressed as molar enthalpy (H_m) which is obtained on dividing the enthalpy by the number of moles. Since, U and V are state properties and P is a state variable, H is also a state variable. Hence, enthalpy change (dH) for any cyclic process is zero. Secondly, enthalpy is an extensive property as it depends on the amount of the system.

On differentiation of Eq. (2.19), we get:

$$dH = dU + PdV + VdP \tag{2.43}$$

or

$$dH = dU + PdV \text{ at constant pressure} \tag{2.44}$$

Now, according to first law of thermodynamics, $dU = \delta q - PdV$. On inserting the value of dU in Eq. (2.44), we obtain:

$$dH = \delta q - PdV + PdV = \delta q \tag{2.45}$$

Hence, for a finite change at constant pressure,

$$q = \Delta H \tag{2.46}$$

Therefore, many textbooks (especially the older ones) refer to enthalpy as heat content of the substance due to the above equality. However, this is not scientifically correct because heat is not a state property.

Now, Eq. (2.43) can be written as:

$$dH = dU + d(PV) \tag{2.47}$$

On dividing both sides of this equation by dP and considering temperature to be constant, we have:

$$\left(\frac{\delta H}{\delta P}\right)_T = \left(\frac{\delta U}{\delta P}\right)_T + \left(\frac{\delta (PV)}{\delta P}\right)_T \tag{2.48}$$

For one mole of an ideal gas at constant temperature, $PV = RT =$ Constant. Hence, d(PV) = 0. In addition, it has been mention in Sect. 2.1, that for a fixed mass, the internal energy of an ideal gas is a function of temperature only and under isothermal condition (i.e. dT = 0), dU = 0. From all these, Eq. (2.48) becomes

$$\left(\frac{\delta H}{\delta P}\right)_T = 0 \tag{2.49}$$

In other words, enthalpy of an ideal gas is independent of pressure at constant temperature. Similarly, enthalpy is independent of volume. Hence, H is a function of temperature only for a fixed mass of the substance. H increase with rise of temperature due to increase in molecular activities. This is of great help in calculating the process heat balance, since in most of the metallurgical processes, the gas phase is having approximately ideal behaviour.

2.4.1 Enthalpy Changes

The absolute value of enthalpy of a system cannot be measured directly. Only a change in enthalpy can be measured. For processes under constant pressure condition, change in enthalpy (ΔH) is equal to the change in internal energy of the system plus work done by the system on its surroundings (i.e. $\Delta H = \Delta U + P\Delta V$) or ΔH is equal to the heat absorbed by the system (i.e. $\Delta H = q$). This constitutes the basis for thermal energy calculations and process heat balance, which is very important for analysis of processes occurring in manufacturing industries. Energy is costly, and there is a continuous attempt to save energy in all types of activity. Without process heat balance, energy audit is not possible. Further, Eq. (2.46) allows experimental determination of ΔH of a process by measurement of q by a calorimeter.

The constant pressure restriction is mostly not important. Energies of solids and liquids are hardly affected by some changes in pressure due to their very small molar volumes. In other words, VdP term in Eq. (2.43) is negligible. In majority of the metallurgical processes, gases are ideal and maximum pressure is of few atmospheres. Therefore, $\Delta H = q$ approximation is quite all right.

2.4.2 Classification of Enthalpy Changes

As per universal convention, enthalpy changes are expressed per mole of a substance. Since enthalpy is a state property, ΔH depends only on initial and final states and not on path followed. For a reversible isothermal process, the temperature remains constant all through. If the process is irreversible and isothermal, then temperature at beginning and end of the process will be the same, but the temperature can vary significantly in between (as shown in Fig. 2.1). For calculation of enthalpy changes in isothermal processes (reversible or irreversible), only the initial and final temperatures are assumed to be the same. Enthalpy changes are classified as follows:

Heat of Reaction: Chemical reactions are carried out either to obtain a useful product or to get chemical energy. By convention, the reaction is considered to be isothermal and isobaric. The change in enthalpy for a chemical reaction (ΔH_r) can be calculated by taking difference of the total heat contents (enthalpies) of the products and reactants.

$$\Delta H_r = \sum H_{\text{products}} - \sum H_{\text{reactants}} \tag{2.50}$$

ΔH_r under standard state (i.e. at 298 K and 1 atm. pressure) is expressed as:

$$\Delta H_r^0 = \sum H_{\text{products}}^0 - \sum H_{\text{reactants}}^0 \tag{2.51}$$

Chemical reactions occurring with the absorption of heat are known as endothermic reactions, and in this case, ΔH_r is positive (since $\sum H_{\text{products}} > \sum H_{\text{reactants}}$). Conversely, heat is evolved for exothermic reactions and ΔH_r is negative (since $\sum H_{\text{products}} < \sum H_{\text{reactants}}$). As shown in Fig. 2.2, exothermicity of the reaction evolves some amount of heat energy and as a result, temperature of the process increases.

Depending upon the type of reaction, the enthalpy change is termed as (i) heat of formation of compound, (ii) heat of decomposition of compound and (iii) heat of combustion. The enthalpy change in the formation of one mole of the compound from its constituent elements at constant temperature and pressure is called heat of formation of compound. Similarly, the heat absorbed or evolved when one mole of a compound decomposes into its constituent elements is referred to as heat of decomposition. In fuel technology, one very frequently uses the term heat of combustion. This

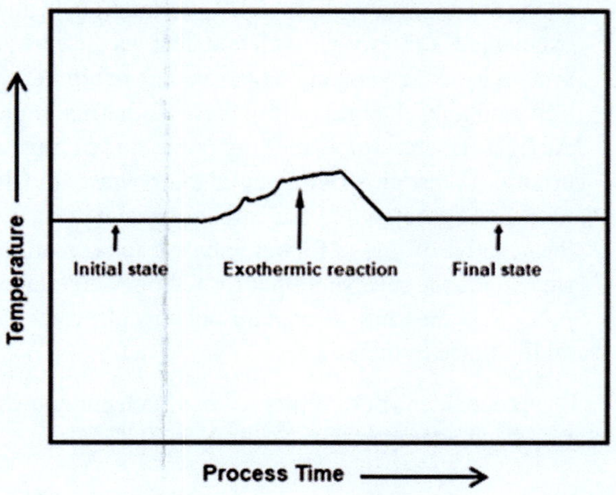

Fig. 2.2 A schematic temperature versus process time plot for an irreversible isothermal reaction

is the amount of heat evolved when one mole of a fuel substance is completely burnt in oxygen to reach up to the level of highest oxidation state. For example, complete combustion of carbon leads to the formation of CO_2 gas as it is the highest oxidation state of carbon. The enthalpy changes in all these are calculated by the formula cited in Eq. (2.50). In order to compare the enthalpies of different substances, it is necessary to select a reference or standard state and this is usually chosen as their stable forms exiting at atmospheric temperature (298 K) and pressure. By convention, the heat contents of the elements are arbitrarily taken as zero when these are in their reference (i.e. standard) states. At the reference temperature, the enthalpy of a compound is equal to its heat of formation.

Sensible Heat/Sensible Enthalpy: It is nothing but the change in enthalpy due to change in temperature of a substance. It has been subdivided as follows:

i Enthalpy Change Without any Change in Physical State of the Material: This has already been discussed in Sect. 2.4.1. As per universal convention, 298 K (i.e. 25 °C) is selected as the reference temperature for calculation of enthalpy at any temperature T and the sensible heat of a substance at this reference temperature is taken as zero. The enthalpy change (per mole) in heating the material from 298 K to a temperature of T is given by the following equation

$$H_T - H_{298} = \int_{298}^{T} C_P \mathrm{d}T \tag{2.52}$$

where C_P is the molar heat capacity at constant pressure.

ii Enthalpy Change Accompanying a Change in Physical State of Material: It is defined as the enthalpy change when one mole of the substance undergoes a specific physical change, i.e. melting, evaporation, condensation, allotropic/phase transformation, etc. The physical state and condition of the substance are dependent on temperature. For example, a pure substance changes from solid to liquid when heated at the melting point and from liquid to gas at the boiling point. The change from solid to gas is called sublimation. In all these changes, the heat energy exchanged is referred to as latent heat because its addition or extraction from the system is not accompanied by a change in temperature of the system. By convention, enthalpy changes during these isothermal transformations are designated by ΔH_m,ΔH_v, etc. Below melting point, one or more changes in the atomic configuration (allotropic or other solid-state phase transformations) may also occur. It may also be noted that the enthalpy change of a reverse phase transformation is the negative of ΔH of forward phase transformation. For example, the enthalpy of freezing is the reverse of that for fusion. The total enthalpy change in a complete process is the sum of enthalpy changes for each of the phase transformations taking place in the process.

Let us consider a pure substance X, which undergoes the following changes during heating from a temperature of 298 K to T K:

$$X(\text{solid}) \quad \text{at } 298\,\text{K} \rightarrow X(\text{liquid}) \text{ at } T_m \rightarrow X(\text{gas}) \text{ at } T_b \rightarrow X(\text{gas}) \text{ at } T$$

where T_m and T_b are the melting and boiling points of X. Now,

$$H_T - H_{298} = \int_{298}^{T_m} C_P(s)\text{d}T + \Delta H_m + \int_{T_m}^{T_b} C_p(l)\text{d}T + \Delta H_V + \int_{T_b}^{T} C_P(g)\text{d}T \quad (2.53)$$

where $C_P(s)$, $C_P(l)$, $C_P(g)$ are molar heat capacities at constant pressure for solid, liquid and gas phases, respectively. ΔH_m and ΔH_v are latent heats of fusion and boiling per mole of substance X. If the substance is not pure, it will not have a single melting or boiling point i.e. it melts and boils over a range of temperature, and in this case, Eq. (2.53) will not be applicable. Sometimes, a substance undergoes a solid-state phase transformation within the considered temperature range. In this case, the enthalpy changes associated with these phase transformations must be added in the above equation.

Let us take an example of allotropic phase transformations in iron, as shown in Table 2.1. Iron exhibits the following structures in the solid state:

δ-Fe melts at 1809 K. Therefore, the enthalpy change during heating of iron from 298 to 2000 K is given as follows:

$$\begin{aligned} H_{2000} - H_{298} = & \int_{298}^{1033} C_P(\alpha)\text{d}T + \Delta H_{\text{tr}}(\alpha \rightarrow \beta)_{1033\text{K}} \\ & + \int_{1033}^{1186} C_P(\beta)\text{d}T + \Delta H_{\text{tr}}(\beta \rightarrow \gamma)_{1186\text{K}} \\ & + \int_{1186}^{1665} C_P(\gamma)\text{d}T + \Delta H_{\text{tr}}(\gamma \rightarrow \delta)_{1665} \\ & + \int_{1665}^{1809} C_P(\delta)\text{d}T + \Delta H_m + \int_{1809}^{2000} C_P(l)\text{d}T \end{aligned} \quad (2.54)$$

Table 2.1 Structures of iron over a range of temperature

Phase	Structure	Temperature range (K) of stability
α	Body-centred cubic (BCC)	Up to 1033
β	Non-magnetic (BCC)	1033–1186
γ	Face-centred cubic (FCC)	1186–1665
δ	Body-centred cubic (BCC)	1665–1809

where $C_P(\alpha)$,$C_P(\beta)$, $C_P(\gamma)$ and $C_P(\delta)$ are the molar heat capacities at constant pressure for α, β, γ and δ phases respectively. The symbol ΔH_{tr} represents enthalpy change during solid-state phase transformation.

Heat of Solution or Mixing: When solids or gases are dissolved in liquids or when two different liquids are mixed, heat is evolved or absorbed. This thermal change per mole of the solute is called the heat of solution. In other words, the process of dissolution of a substance in a solvent is accompanied by a change in enthalpy (ΔH_{mix}).

Let components 1, 2, ... n are mixed together to form a solution. Now, the total enthalpy change of mixing is given by the following equation:

$$\Delta H_{mix} = n_1 \Delta H_1 + n_2 \Delta H_2 + \cdots + n_n \Delta H_n \quad (2.55)$$

where ΔH_1, ΔH_2, ... ΔH_n are the change in partial molar enthalpies of components 1, 2, ... n; n_1, n_2,... n_n are the number of moles of these components. The thermodynamics involved in the formation of solutions has been discussed in detailed in Chap. 6.

2.5 Thermochemical Laws

Thermochemistry is the study of heat exchanges during chemical or physico-chemical processes. The heat exchanged in chemical or metallurgical processes is seen to be a path dependent property. These processes are usually carried out under isobaric or isochoric conditions, and the observations derived are as follows:

$$dU = \delta q(\text{for isochoric processes}) \text{ and } dH = \delta q(\text{for isochoric processes})$$

U and H are state functions. Hence, the heat exchanges in chemical and metallurgical processes, in general, are not path dependent but state functions. Based on this fact, two thermochemical laws were developed: (i) Hess's law and (ii) Kirchhoff's law. First major break-through in the formulation of thermochemical laws was achieved by G. H. Hess which was followed by Kirchhoff's studies on the effect of temperature on the enthalpy of the reaction. These laws originally proposed for the calculation of heat of reaction (ΔH_r) are discussed in the following sections.

2.5.1 Hess's Law

This law was developed by G. H. Hess in 1840, even before the formulation of first law of thermodynamics. But, it is now considered as a corollary of this first law. He suggested that the total heat exchanged in a chemical process remains the same irrespective of whether the process occurs in a single stage or through multiple stages.

Alternatively, Hess's law states that for an isothermal process, the enthalpy change for a reaction remains the same whether it takes place in one or several steps, i.e. the heat of reaction (ΔH_r) depends only on the initial and final states of the process. This law can be explained by an illustration of reduction of iron oxide (Fe_2O_3) by carbon as follows:

The reduction of iron oxide by carbon can be carried out in one stage as:

$$Fe_2O_3 + 3C = 2Fe + 3CO \tag{2.56}$$

Alternatively in multiple stages as:

$$3Fe_2O_3 + C = 2Fe_3O_4 + CO \tag{2.57}$$

$$2Fe_3O_4 + 2C = 6FeO + 2CO \tag{2.58}$$

$$6FeO + 6C = 6Fe + 6CO \tag{2.59}$$

ΔH for single stage reduction, i.e. reaction (2.56), is given by

$$\Delta H_{(2.56)} = 2H_{Fe} + 3H_{CO} - H_{Fe_2O_3} - 3H_C \tag{2.60}$$

And expressions for ΔH in multistage reduction, i.e. reactions (2.57) to (2.59), are given by

$$\Delta H_{(2.57)} = 2H_{Fe_3O_4} + 3H_{CO} - 3H_{Fe_2O_3} - 3H_C \tag{2.61}$$

$$\Delta H_{(2.58)} = 6H_{FeO} + 2H_{CO} - 2H_{Fe_3O_4} - 2H_C \tag{2.62}$$

$$\Delta H_{(2.59)} = 6H_{Fe} + 6H_{CO} - 6H_{FeO} - 6H_C \tag{2.63}$$

The addition of Eqs. (2.61) to (2.63) gives the same equation as in Eq. (2.60).

Applications: Followings are the innumerable applications of Hess's law in the fields of chemical and metallurgical engineering:

(i) This law is helpful in calculating the enthalpy changes/heats of many reactions from those of other associated reactions occurring at the same temperature.

(ii) An useful practical application of this law is to calculate the enthalpy changes/heats of those reactions which cannot be determined experimentally because of their sluggishness or experimental difficulties.

(iii) Hess's law is also useful in calculating the heat exchanged/enthalpy change during change in structure of a solid, solid-state phase transformation, allotropic modification, etc.

(iv) For the application of Hess's law, initial and final temperatures during the process should be the same. It does not matter whatever the path is followed in between.

2.5.2 *Kirchhoff's Law*

From Eq. (2.30), we know that $C_P = \left(\frac{\delta H}{\delta T}\right)_P = \frac{\mathrm{d}H}{\mathrm{d}T}$. For reaction at constant pressure, this equation can be written as follows:

$$\Delta C_P = \left(\frac{\delta \Delta H}{\delta T}\right)_P = \frac{\mathrm{d}\Delta H}{\mathrm{d}T} \tag{2.64}$$

This equation means that the rate of change in the value of ΔH of a process/reaction with temperature is equal to the change in molar/specific heat capacity (i.e. ΔC_P). This is known as Kirchhoff's equation. Equation (2.64) can be rearranged as:

$$\mathrm{d}(\Delta H) = \Delta C_P \mathrm{d}T \tag{2.65}$$

Let the temperature of the substance/system changes from T_1 to T_2. Now, on integration of Eq. (2.65) over the temperature range from T_1 to T_2, we get the following:

$$\int_{\Delta H_1}^{\Delta H_2} \mathrm{d}(\Delta H) = \int_{T_1}^{T_2} \Delta C_P \mathrm{d}T \text{ Or } \Delta H_2 - \Delta H_1 = \int_{T_1}^{T_2} \Delta C_P \mathrm{d}T \tag{2.66}$$

where ΔH_2 and ΔH_1 are the heats of reaction (i.e. change in enthalpy) at temperatures T_1 and T_2 under isobaric condition, and ΔC_P is the change in molar/specific heat capacity at constant pressure for change of temperature from T_1 to T_2. Equation (2.66) is the integrated form of Kirchhoff's equation. As is clear from this equation, heat of reaction/enthalpy change at temperature T_2 (i.e. ΔH_2) could be found out if heat of reaction/enthalpy change at temperature T_1 (i.e. ΔH_1) and the value of ΔC_P over a temperature range from T_1 to T_2 are known. We also know that $\Delta C_P = \sum C_{P,\text{products}} - \sum C_{P,\text{reatants}}$. Hence, Eq. (2.66) becomes

$$\Delta H_2 = \Delta H_1 + \int_{T_1}^{T_2} \left(\sum C_{P,\text{products}} - \sum C_{P,\text{reactants}}\right) \mathrm{d}T \tag{2.67}$$

This equation is another form of Kirchhoff's equation. Thus, it is clear from Eq. (2.67) that change in heat of reaction with temperature is directly related to the molar heat capacities of reactants and the products. Hence, according to Kirchhoff's law, the heat of reaction is a function of temperature. If the reactants and the products are pure and in their standard states, Eq. (2.66) can be written as follows:

$$\Delta H_2^o = \Delta H_1^o + \int_{T_1}^{T_2} \Delta C_P^o \mathrm{d}T \quad (2.68)$$

In case of melting, vapourization and any other phase transformations, Eq. (2.67) gets simplified. For example, in case of melting of solid at temperature T_2, this equation becomes:

$$\Delta H_m(\text{at } T_2) = \Delta H_1 + \int_{T_1}^{T_2} [C_P(l) - C_P(s)]\mathrm{d}T \quad (2.69)$$

2.6 Internal Energy Versus Enthalpy

For isobaric processes, equation $H = U + PV$ leads to the following expression:

$$\Delta H = \Delta U + P\Delta V \quad (2.70)$$

The chemical processes involving only solids and liquids (e.g. $2CaO{\cdot}SiO_2 = 2CaO + SiO_2$), the change in volume is generally very small and hence, the term $P\Delta V$ (i.e. capacity of the system to do work) can be neglected in comparison to ΔU. For these processes, the change in enthalpy of the system is almost equal to the change in its internal energy (i.e. $\Delta H \approx \Delta U$). Application of first law of thermodynamicsd$U = \delta q - P\mathrm{d}V$) in such processes also clarifies that the entire amount of heat energy supplied to the system is totally utilized in increasing the internal energy of the system (i.e. $q = \Delta U$). The difference between U and H arises because of the presence of gaseous component in the system. The involvement of gaseous component in the system gives perceptible value for $P\Delta V$ and hence, ΔH is distinctly different from ΔU (i.e. $\Delta H \neq \Delta U$).

Let us take the example of calcination of limestone by the following reaction:

$$CaCO_3 = CaO + CO_2 \quad (2.71)$$

There is a tremendous increase in the volume of the system on calcination due to the evolution of carbon dioxide gas and its expansion in the system does $P\Delta V$ amount of work on the surrounding. As a result, ΔH becomes > ΔU (i.e. $\Delta H > \Delta U$). Hence, in addition to the requirement of energy for breaking the chemical bond between CaO and CO_2 in limestone, an additional energy equivalent to $P\Delta V$

making a total of $\Delta U + P\Delta V$ needs to be supplied for the process to occur and this is the enthalpy change of the system on calcination. In short, the supplied heat energy gets totally utilized in increasing the enthalpy of the system (i.e. $q = \Delta H$).

Solved Problems

Problem 2.1 Due to expansion, the pressure of 10 L of an ideal gas at a temperature of 25 °C drops from 10 to 1 atm. Given that C_V of the gas = 1.5 R. Find out the values of the amount of work done, amount of heat exchanged, change in internal energy (ΔU) and change in enthalpy (ΔH), if the process occurs under (a) isothermal and reversible conditions and (b) adiabatic and reversible conditions.

Solution

We know that $PV = nRT$.

or

$$\begin{aligned} n(\text{No. of moles of gas}) &= PV/RT \\ &= (10 \times 10)/(0.082 \times 298) \text{ as } R = 0.082\, L \cdot \text{atm/K mol} \\ &= 4.09 \end{aligned}$$

(a) For isothermal and reversible process: Let the gas moves from a state 1 to state 2. Then, $P_1V_1 = P_2V_2$ or $V_2 = \frac{P_1V_1}{P_2} = (10 \times 10)/1 = 100\,\text{L}$.

As is known to us, $\Delta U = 0$ for an isothermal process. From the first law of thermodynamics,

$$dU = \delta q - \delta w = 0 \quad \text{or} \quad \delta q = \delta w.$$

where δq is the amount of heat exchanged.

Now, on integration between the limits,

$$\begin{aligned} q &= \int_1^2 P\text{d}V = n\text{RT}\int_1^2 \frac{\text{d}V}{V} \\ &= 4.09 \times 8.314 \times 298 \times \ln(100/10) \\ &= 23.33\,\text{kJ}. \end{aligned}$$

Hence, the amount of work done = 23.33 kJ as $q = w$.

Now, $\Delta H = \Delta U + P_2V_2 - P_1V_1 = 0 + P_2V_2 - P_1V = nR(T_2 - T_1) = 0$, as for an isothermal process$T_1 = T_2$.

(b) For reversible and adiabatic process: For reversible adiabatic process, $P_1V_1^\gamma = P_2V_2^\gamma$ or $10 \times 10^\gamma = 1 \times V_2^\gamma$ or $V_2^\gamma = 10 \times 10^\gamma$

We know that $C_P - C_V = R$ and $\gamma = C_P/C_V$. Hence, $R/C_V = \gamma - 1$
Given that $C_V = 1.5R$. Hence, $R/1.5R = \gamma - 1$ or $\gamma = 5/3$
On inserting the value of γ in the above equation, we get

$$V_2^{5/3} = 10 \times 10^{5/3}$$

or

$$V_2 = \left(10 \times 10^{5/3}\right)^{3/5} = 40\text{L}$$

Now,

$$T_2 = \frac{P_2 V_2}{nR} = \frac{1 \times 40}{4.09 \times 0.082} = 119.27\text{K}$$

From first law of thermodynamics for an adiabatic process $\mathrm{d}U = -\delta w$.

We also know that $\mathrm{d}U = nC_V\mathrm{d}T$. Hence, the above equation becomes $\mathrm{d}U = nC_V\mathrm{d}T = -\delta w$ or $-\delta w = nC_V dT$.

On integration, we get

$$-w = \int_{T_1}^{T_2} nC_V\mathrm{d}T = nC_V(T_2 - T_1)$$
$$= 4.09 \times 1.5 \times 8.314(119.27 - 298) = -9.12$$
$$w = 9.12\,\text{kJ}.$$

Hence, $\Delta U = -w = -9.12\,\text{kJ}$.
Since, the process is adiabatic, the amount of heat exchanged = 0.
Now,

$$\Delta H = \Delta U + P_2V_2 - P_1V_1 = \Delta U + nR(T_2 - T_1)$$
$$= -9.12 + 4.09 \times 8.314 \times (119.27 - 298)$$
$$= -6.09\,\text{kJ/mol}.$$

Problem 2.2 Find out the value of ΔH^o for the reaction Cu_2S (s) + $2Cu_2O$ (s) = 6Cu (l) + SO_2 (g) at the temperature of 1250 °C. Followings are the data for standard enthalpy changes (ΔH^o) of formation at the above temperature.

ΔH^o for Cu_2S (s) = – 86.7 kJ/mol., ΔH^o for Cu_2O (s) = – 176.4 kJ/mol., ΔH^o for Cu (l) = 0 and ΔH^o for SO_2 (g) = – 278.4 kJ/mol.

Solution
We know that

$$\Delta H^o = \Sigma \Delta H^o(\text{products}) - \Sigma \Delta H^o(\text{reactants})$$

Hence,

$$\begin{aligned}\Delta H^o_{1523} &= \left[\left(6\Delta H^o_{\text{Cu}} + \Delta H^o_{\text{SO}_2}\right) - \left(\Delta H^o_{\text{Cu}_2\text{S}} + 2\Delta H^o_{\text{Cu}_2\text{O}}\right)\right] \\ &= (0 - 278.4) - (-86.7 - 2 \times 176.4) \\ &= -278.4 + 86.7 + 352.8 = 161.1\ \text{kJ/mol}.\end{aligned}$$

Problem 2.3 Find out the value of ΔH^o in the formation of PbO (s) from Pb (l) and O_2 (g) at 527 °C. Data given are as follows:

$$\text{Melting point of Pb} = 327\,^\circ\text{C}$$

$$\text{Latent heat of melting } (L_m) = 4.81\ \text{kJ/mol}$$

$$\Delta H^o_{\text{PbO},298} = 219.24\ \text{kJ/mol}$$

$$C_P \quad \text{for PbO(s)} = 44.35 + 16.74 \times 10^{-3} T\ \text{J/K mol}$$

$$C_P \quad \text{for Pb(s)} = 23.56 + 9.75 \times 10^{-3} T\ \text{J/K mol}$$

$$C_P \quad \text{for Pb(l)} = 32.43 - 3.10 \times 10^{-3} T\ \text{J/K mol}$$

$$C_P \quad \text{for O}_2\text{(g)} = 29.96 + 4.184 \times 10^{-3} T - 1.67 \times 10^5\ \text{J/K mol}$$

Solution

The chemical reaction for this problem is: Pb (l) + 0.5 O_2 (g) = PbO (s).

Total standard enthalpy change

$$\begin{aligned}\left(\Delta H^o_t\right) &= \Delta H^o_{800} - \Delta H^o_{298} \\ &= \int_{298}^{600} C_P \text{Pb(s)} \mathrm{d}T + L_m + \int_{600}^{800} \Delta C_P \mathrm{d}T\end{aligned}$$

Calculation of each term separately yields

$$\int_{298}^{600} C_P \text{Pb}(s) \mathrm{d}T = \int_{298}^{600} \left(23.56 + 9.75 \times 10^{-3} T\right) \mathrm{d}T$$

$$= 23.56(600 - 298) + \frac{9.75 \times 10^{-3}}{2}\left(600^2 - 298^2\right)$$
$$= 23.56 \times 302 + 4.875 \times 10^{-3} \times 271{,}196$$
$$= 8437.2\ \mathrm{J/mol}$$

$$\int_{600}^{800} \Delta C_P \mathrm{d}T = \int_{600}^{800} [\Sigma C_P(\text{products}) - \Sigma C_P(\text{reactants})]\mathrm{d}T$$
$$= \int_{600}^{800} [(44.35 + 16.74 \times 10^{-3} T) - (32.43 - 3.10 \times 10^{-3} T)$$
$$- 0.5(29.96 + 4.184 \times 10^{-3} T - 1.67 \times 10^5)]\mathrm{d}T$$
$$= -3.06(800 - 600) + 8.874 \times 10^{-3}\left(800^2 - 600^2\right)$$
$$+ 0.835(800 - 600) = 2039.7\ \mathrm{J/mol}$$

$$L_m = 4.81\ \mathrm{kJ/mol} = 4810\ \mathrm{J/mol}$$

Addition of all these terms yields the following:

$$\Delta H_t^o = 8437.2 + 4810 + 2039.7 = 15{,}286.9\ \mathrm{J/mol} = 15.29\ \mathrm{kJ/mol}$$

Hence,

$$\Delta H_{800}^o = \Delta H_{298}^o + 15.29 = -219.24 + 15.29 = -204\ \mathrm{kJ/mol}$$

Problem 2.4 By calculation, determine the value of ΔH^o at 25 °C for the following reaction

$$\mathrm{PbS(s)} + 1.5\mathrm{O_2(g)} = \mathrm{PbO(s)} + \mathrm{SO_2(g)}$$

Data given are as follows:

$$\Delta H_{298}^o \text{ for PbS(s)} = -22.61\ \mathrm{kcal/mol},$$
$$\Delta H_{298}^o \text{ for PbO(s)} = -52.75\ \mathrm{kcal/mol} \text{ and}$$
$$\Delta H_{298}^o \text{ for } \mathrm{SO_2(g)} = -71.29\ \mathrm{kcal/mol}$$

Solution
We know that

$$\Delta H^o = \Sigma \Delta H^o(\text{products}) - \Sigma \Delta H^o(\text{reactants})$$

$$= \left[\Delta H^o_{298}\text{PbO(s)} + \Delta H^o_{298}\text{SO}_2\text{(g)}\right] - \left[\Delta H^o_{298}\text{PbS(s)} + 1.5\Delta H^o_{298}\text{O}_2\text{(g)}\right]$$
$$= (-52.75 - 71.29) - (-22.61 + 1.5 \times 0)$$
$$= -101.43\,\text{kcal/mol}$$

Problem 2.5 Determine the value of standard enthalpy change for the reaction CaO (s) + SiO_2 (s) = CaO·SiO_2 (s) at 25 °C. The values of ΔH^o associated with the formation of CaO, SiO_2 and CaO·SiO_2 from their components at the above temperature are – 149.52, – 216.6 and – 53.18 kcal/mol.

Solution
We know that

$$\Delta H^o = \Sigma\Delta H^o(\text{products}) - \Sigma\Delta H^o(\text{reactants})$$
$$= \Delta H^o_{298}\text{CaO}\cdot\text{SiO}_2\text{(s)} - \left[\Delta H^o_{298}\text{CaO(s)} + \Delta H^o_{298}\text{SiO}_2\text{(s)}\right]$$
$$= -53.18 + 149.52 + 216.6 = 312.94\,\text{kcal/mol}$$

Problem 2.6 Determine the value of ΔH^o at 1 atm. and 25 °C for the reaction 3FeO (s) + 2Al (s) = Al_2O_3 (s) + 3Fe (s) in terms of the followings:

(a) Per mole of Fe produced, (b) Per mole of Al_2O_3 produced, (c) Per mole of reacted Al, (d) Per mole of reacted FeO and (e) Per gram of Fe produced. Data given are as follows:

$$\Delta H^o_{298} \text{ for FeO} = -63.36\,\text{kcal/mol and } \Delta H^o_{298} \text{ for Al}_2\text{O}_3 = -400.38\,\text{kcal/mol}$$

Solution
We know that

$$\Delta H^o = \Sigma\Delta H^o(\text{products}) - \Sigma\Delta H^o(\text{reactants})$$
$$= \left[\Delta H^o_{298}\text{AbO}_3\text{(s)} + 3\Delta H^o_{298}\text{Fe(s)}\right] - \left[3\Delta H^o_{298}\text{FeO(s)} + 2\Delta H^o_{298}\text{Al(s)}\right]$$
$$= (-400.38 + 3 \times 0) - (-3 \times 63.36 + 2 \times 0)$$
$$= -210.3\,\text{kcal/mol}$$

(a) ΔH^o_{298} per mole of Fe produced $= -210.3/3 = -70.1$ kcal
(b) ΔH^o_{298} per mole of Al_2O_3 produced = – 210.3 kcal
(c) ΔH^o_{298} per mole of reacted Al = – 210.3/2 = – 105.15 kcal
(d) ΔH^o_{298} per mole of reacted FeO = t 210.3/3 = t 70.1 kcal
(e) ΔH^o_{298} per gram of Fe produced = ΔH^o_{298} per gram of Fe produced/Atomic Weight of Fe = 70.1/56 = 1.25 kcal

Problem 2.7 Find out the value of ΔH^o for the reaction CO (g) + 0.5 O_2 (g) = CO_2 (g) at 200 °C. Given that the values of ΔH^o associated with the formation of CO and CO_2 gases at 298 K are – 110.5 kJ/mol and – 393.5 kJ/mol. Followings are the values of their molar heat capacities:

$$C_P \text{ for CO(g)} = 30.0 + 0.0041T \text{ J/K mol}$$

$$C_P \text{ for CO(g)} = 44.2 + 0.0088T \text{ J/K mol}$$

$$C_P \text{ for O}_2\text{(g)} = 28.5 + 0.0042T \text{ J/K mol}$$

Solution

We know that

$$\begin{aligned}\Delta H^o &= \sum \Delta H^o_{\text{(products)}} - \sum \Delta H^o_{\text{(reactants)}} \\ &= \Delta H^o_{298}\text{CO}_2(\text{g}) - \left[\Delta H^o_{298}\text{CO}_2(\text{g}) + 0.5\Delta H^o_{298}\text{O}_2(\text{g})\right] \\ &= \left[-39.35 \times 10^3 - \left(-110.5 \times 10^3 + 0\right)\right] \\ &= -283 \times 10^3 \text{ J/mol}\end{aligned}$$

Now,

$$\begin{aligned}\Delta C_P &= \Sigma C_P(\text{products}) - \Sigma C_P(\text{reactants}) \\ &= \left[C_P\text{CO}_2(\text{g}) - \{C_P\text{CO(g)} + 0.5C_P\text{O}_2\,(\text{g})\}\right] \\ &= [(44.2 \times 0.0088T) - \{(30.0 + 0.0041T) + 0.5 \times (28.5 + 0.042T)\}] \\ &= [(44.2 - 30.0 - 14.25) + (0.0088T - 0.0041T - 0.0021T)] \\ &= -0.05 + 0.0026T\end{aligned}$$

As is known to us

$$\Delta H^o_t = \Delta H^o_{473} - \Delta H^o_{298} = \int_{298}^{473} \Delta C_P \text{d}T$$

$$\begin{aligned}\Delta H^o_{473} &= \Delta H^o_{298} + \int_{298}^{473} \Delta C_P \text{d}T \\ &= -283 \times 10^3 + \int_{298}^{473} (-0.05 + 0.0026T)\text{d}T \\ &= -283 \times 10^3 - 0.05\,(473 - 298) + 0.0026/2\left(473^2 - 298^2\right)\end{aligned}$$

$$= 283 \times 10^3 - 8.75 + 17.54 = -282{,}283.35\ \text{J/mol}$$
$$= -28.283\ \text{kJ/mol}$$

Problem 2.8 Determine the value of standard enthalpy change (ΔH^o) involved in the formation of WO_3 (s) from tungsten (W) and oxygen gas at 25 °C and 1 atm. pressure with the help the following data:

$$W(s) + O_2(g) = WO_2(s) \quad \Delta H^o_{298} = -134.13\,\text{kcal/mol} \tag{2.72a}$$

$$3WO_2(s) + O_2(g) = W_3O_8(s) \quad \Delta H^o_{298} = -131.63\,\text{kcal/mol} \tag{2.73b}$$

$$W_3O_8(s) + 0.5O_2(g) = 3WO_3(s) \quad \Delta H^o_{298} = -22.19\,\text{kca/mol} \tag{2.74c}$$

Solution
Equation (2.72a) can be arranged as follows:

$$3W(s) + 3O_2(g) = 3WO_2(s) \quad \Delta H^o_{298} = -402.39\ \text{kcal/mol} \tag{2.75d}$$

On addition of Eqs. (2.72b), (2.72c) and (2.72d), we get the following:

$$3W(s) + 4.5O_2(g) = 3WO_3(s) \quad \Delta H^o_{298} = -[131.63 + 22.19 + 402.39]$$
$$= -556.21\ \text{kcal/mol}$$

or

$$W(s) + 1.5O_2(g) = WO_3(s) \quad \Delta H^o_{298} = -556.21/3 = -185.40\ \text{kcal/mol}$$

Problem 2.9 Metallothermic reduction of hematite (Fe_2O_3) by aluminium (Al) occurs by the reaction $Fe_2O_3 + 2Al = 2Fe + Al_2O_3$. Find out the followings:

(i) How much amount of heat is generated during this metallothermic reduction reaction?
(ii) How much amount of Al will be required to produce 1 kg of Fe.

Data given are as follows:

ΔH^o for Fe_2O_3 = – 196.49 kcal/mol and ΔH^o for Al_2O_3 = – 401.28 kcal/mol

Solution
The chemical reaction is $Fe_2O_3 + 2Al = 2Fe + Al_2O_3$.

Molecular weight of Fe_2O_3 = 160 kg.

Molecular weight of $Al_2O_3 = 102$ kg.
We know that

$$\begin{aligned}\Delta H^o &= \sum \Delta H^o(\text{products}) - \sum \Delta H^o(\text{reactants}) \\ &= [2 \times \Delta H^o\text{Fe} + \Delta H^o\text{Al}_2\text{O}_3] - [\Delta H^o\text{Fe}_2\text{O}_3 + 2 \times \Delta H^o\text{Al}] \\ &= [0 - 401.28] - [-196.49 + 0] = -204.79 \text{ kcal/mol}\end{aligned}$$

Hence, heat generated during the above metallothermic reduction reaction = − 204.79 kcal/mol.

Now, for 2 × 56 kg of Fe production, Al required = 2 × 27 = 54 kg.
For 1 kg of Fe production, Al required = 54/112 = 0.48 kg.

Problem 2.10 Find out the value of standard enthalpy change (ΔH^o) for the metallothermic reduction of Cr_2O_3 by Al at 627 °C. Data given are as follows:

$$\Delta H^o_{298} \text{ for } \text{Al}_2\text{O}_3(\text{s}) = -1674.0 \text{ kJ/mol and } \Delta H^o_{298} \text{ for } \text{Cr}_2\text{O}_3(\text{s}) = -1120.3 \text{ kJ/mol}$$

$$C_P \text{ for Al}_2\text{O}_3(\text{s}) = 106.6 + 17.8 \times 10^{-3}\text{T} - 28.5 \times 10^5\text{T}^{-2} \text{ J/Kmol}$$

$$C_P \text{ for Al(s)} = 20.7 + 12.4 \times 10^{-3} \text{ T J/Kmol}$$

$$C_P \text{ for Cr}_2\text{O}_3(\text{s}) = 119.4 + 9.2 \times 10^{-3} \text{ T} - 15.6 \times 10^5 \text{ T}^{-2} \text{ J/Kmol}$$

$$C_P \text{ for Cr(s)} = 24.4 + 9.87 \times 10^{-3} \text{ T} - 3.7 \times 10^5 \text{ T}^{-2} \text{ J/Kmol}$$

Solution

The chemical reaction is Cr_2O_3 (s) + 2Al (s) = 2Cr (s) + Al_2O_3 (s).
We know that

$$\begin{aligned}\Delta C_P &= \Sigma C_P(\text{products}) - \Sigma C_P(\text{reactants}) \\ &= [2 \times C_P\text{Cr(s)} + C_P\text{Al}_2\text{O}_3(\text{s})] - [C_P\text{Cr}_2\text{O}_3(\text{s}) + 2 \times C_P\text{Al(s)}] \\ &= [2\left(24.4 + 9.87 \times 10^{-3}\text{T} - 3.7 \times 10^5\text{T}^{-2}\right) \\ &\quad + \left(106.6 + 17.8 \times 10^{-3}\text{T} - 28.5 \times 10^5\text{T}^{-2}\right)] \\ &\quad - \left[\left(119.4 + 9.2 \times 10^{-3}\text{T} - 15.6 \times 10^5\text{T}^{-2}\right) + 2\left(20.7 + 12.4 \times 10^{-3}\text{T}\right)\right] \\ &= (48.8 + 106.6 - 119.4 - 41.4) + 10^{-3}\text{T}(19.74 + 17.8 - 9.2 - 24.8) \\ &\quad - 10^5\text{T}^{-2}(7.4 + 28.5 - 15.6) \\ &= -5.4 + 3.54 \times 10^{-3}\text{T} - 20.3 \times 10^5\text{T}^{-2}\end{aligned}$$

$$\begin{aligned}\Delta H^o_{298} &= \Sigma\Delta H^o_{298}(\text{products}) - \Sigma\Delta H^o_{298}(\text{reactants})\\ &= \left[2\times\Delta H^o_{298}\text{Cr(s)} + \Delta H^o_{298}\text{Al}_2\text{O}_3\text{(s)}\right]\\ &\quad - \left[\Delta H^o_{298}\text{Cr}_2\text{O}_3\text{(s)} + 2\times\Delta H^o_{298}\text{Al(s)}\right]\\ &= [0 - 1674.0] - [-1120.3 + 0] = -553.7\ \text{kJ/mol}\\ &= -553{,}700\ \text{J/mol}\end{aligned}$$

As is known to us

$$\Delta H^o_t = \Delta H^o_{900} - \Delta H^o_{298} = \int_{298}^{900} \Delta C_P \mathrm{d}T$$

or

$$\begin{aligned}\Delta H^o_{900} &= \Delta H^o_{298} + \int_{298}^{900} \Delta C_P \mathrm{d}T\\ &= -55{,}370 + \int_{298}^{900} \left(-5.4 + 3.54\times10^{-3}T - 20.3\times10^5T^{-2}\right)\mathrm{d}T\\ &= -553{,}700 - 5.4(900-298) + 1.77\times10^{-3}\left(900^2 - 298^2\right)\\ &\quad + 20.3\times10^5\left(\frac{1}{900} - \frac{1}{298}\right)\\ &= -556{,}950.8 + 1276.52 - 4466 = -560{,}140.28\ \text{J/mol}\\ &= -560.14\ \text{kJ/mol}\end{aligned}$$

Problem 2.11 Prove the following:

$$C_P = \left(\frac{\delta U}{\delta T}\right)_P + P\left(\frac{\delta V}{\delta T}\right)_P$$

Solution

We know that

$$H = U + PV$$

On differentiation w.r.t temperature T at constant pressure P, we get

$$\left(\frac{\delta H}{\delta T}\right)_P = \left(\frac{\delta U}{\delta T}\right)_P + P\left(\frac{\delta V}{\delta T}\right)_P$$

As is known,

$$C_P = \left(\frac{\delta H}{\delta T}\right)_P$$

Hence,

$$C_P = \left(\frac{\delta U}{\delta T}\right)_P + P\left(\frac{\delta V}{\delta T}\right)_P$$

Problem 2.12 Determine the value of standard enthalpy change when one mole of water at 298 K and 1 atm. pressure is converted into water vapour at 403 K and 2 atm. Data given are as follows:

C_P for water = 75.3 J/K mol, latent heat of water vapour (L_V) = 40.64 × 10^3 J/mol, C_P for water vapour between 373 and 403 K = 30.11 + 9.937 × 10^{-3} T + 0.837 × 10^{-6}T^2 J/K mol.

Solution

$$\begin{aligned}\Delta H &= \int_{298}^{373} C_P(\text{water})\,\mathrm{d}T + L_V + \int_{373}^{403} C_P(\text{water vapour})\mathrm{d}T \\ &= 75.3(373-298) + 40.64\times 10^3 + 30.11(403-373) \\ &\quad + 4.969\times 10^{-3}\left(403^2 - 373^2\right) + 0.279\times 10^{-6}\left(403^3 - 373^3\right) \\ &= 5647.5 + 40{,}640 + 903.3 + 115.68 + 3.78 = 47{,}310.26\ \text{J/mol} \\ &= 47.31\ \text{kJ/mol}\end{aligned}$$

Exercise

2.1 Find Out the Value of Standard Enthalpy Change (ΔH^o) Involved in the Formation of Ethane (C2H6) at 25 °C with the Help of the Following Data:

$$H_2(g) + 0.5O_2(g) = H_2O(l) \quad \Delta H^o_{298} = -68.42\ \text{kcal/mol}$$

$$C(s) + O_2(g) = CO_2(g) \quad \Delta H^o_{298} = -94.02\ \text{kcal/mol}$$

$$C_2H_6(g) + 3.5O_2(g) = 2CO_2(g) + 3H_2O(l) \quad \Delta H^o_{298} = -373.44\ \text{kcal/mol}$$

[**Ans:** – 83 kJ/mol].

2.2 The values of standard enthalpy change involved in the combustion of H_2 (g), CO (g) and CH_3OH (l) at 25 °C are – 68.42, – 67.70 and – 170.81 kcal/mol. Consider the reaction:

$$CO(g) + 2H_2(g) = CH_3OH(1)$$

[**Ans**: 141 kJ/mol].

Chapter 3
Second Law of Thermodynamics and Entropy

A system undergoing a process needs characterization for its many of the thermodynamic properties. This chapter includes thirteen sections and is primarily concerned with second law of thermodynamics and entropy changes (ΔS). Sections 3.2 to 3.4 and 3.7 present introduction, statements and mathematical formulation of second law of thermodynamics followed by illustration on Carnot's cycle. A thorough discussion on entropy changes associated with various processes like reversible, irreversible/spontaneous, chemical reactions, phase transformation, etc. have been provided in this chapter. In addition to the detailed discussion on ΔS involved in these processes, serious attempts have been made in this chapter to impart a complete knowledge to the readers on change in entropy during processing of materials in solid state (e.g. graphitization of carbon, mechanical working of materials, hardening of steel, etc.). Concept on degradation of energy and an increase in entropy has been highlighted in Sect. 3.10.

Equations for variation of entropy with temperature under isobaric and isochoric conditions have also been obtained and interpreted sufficiently for determination of entropy of a system/material at a particular temperature. This chapter got finished by including a large number of solved problems and several exercises.

3.1 Introduction

First law of thermodynamics states that the energy can change its form and is conserved during any thermodynamic process. However, it provides no information regarding the feasibility and direction of a process under a given set of constraints imposed on the system. One may argue that all the reactions proceed towards the attainment of equilibrium and thus their directions are uniquely defined, but the state of equilibrium and its criteria are not defined. The second law gives an answer to this

S. K. Sahoo et al., *Fundamentals of Metallurgical Thermodynamics*,
https://doi.org/10.1007/978-981-99-6671-4_3

problem and provides criteria to predict whether a particular process/reaction would take place under certain specified conditions or not and thus is of great importance.

Another important aspect of second law of thermodynamics is also concerned with conversion of heat into work. The first law of thermodynamics states that energy can be converted from one form to other. But, it does not tell anything either about the limitation in the conversion of energy or the condition necessary for such a conversion. For example, mechanical work in friction is completely converted into heat, but no heat engine can convert the heat completely into work. In heat engines, the efficiency of conversion of heat energy into work is quiet low, i.e. 10–40%. According to second law, the important cause of this low efficiency is the irreversible nature of heat flow. This means that maximum efficiency is obtained in case of reversible process, and a thermodynamic parameter is necessary to judge the energy changes in such processes. This parameter is entropy which clearly distinguishes between reversible (i.e. equilibrium) and irreversible (i.e. non-equilibrium) processes.

The discovery of second law of thermodynamics is connected to the theoretical studies on the efficiency of heat engines by Sadi Carnot in 1824. Later on Clausius, Kelvin, Thompson and Max Planck provided sound theoretical basis and general statements of this law. As in the case of first law of thermodynamics, this law is also based on experimental facts. Before enunciating the different statements of this law, some day-to-day observations of natural processes and simple experiments, dealing with transformation of energy from one form to another, need to be analysed.

3.2 Observations Leading to the Formulation of Second Law

Observations on the Conversion of Mechanical Work to Heat: When two pieces of metals or stones are rubbed against each other, heat is produced due to development of friction. This production of heat is due to work done on the system. As a result of conversion of mechanical work into heat, the thermodynamic states of the heat receiving bodies (presently pieces of metals or stones) change because of a change in their temperature without affecting the surroundings. Thus, the conversion of work into heat leads only to a change in the state of a heat receiving body.

An adiabatic compression of a gas leads to an increase in its temperature and ultimately its internal energy. Here, one may think that this increase in temperature would have been achieved through the supply of heat to the gas. Hence, it must be further stated here that the work of compression has got converted into heat which is stored as an increased internal energy in the gas. As a result, the thermodynamic state of the gas gets affected. From this experiment, it could be further concluded that the work manifests itself in the form of internal energy under an adiabatic condition. The net conclusion is that the conversion of work into heat cause a change in the thermodynamic state of the substance on which work has been done directly or indirectly without affecting other body/system or surrounding.

Observations on the Conversion of Heat into Mechanical Work: Let us consider a reversible isothermal process with an ideal gas as the working substance, which takes heat from the surroundings and convert it to work by expansion. This process brings about two changes: (i) the heat transfer from the surroundings causes a change in their thermodynamic states, and (ii) conversion of heat into work causes a change in the thermodynamic state of the gas (working substance) due to increase in its volume. Further, the initial state can only be restored by compression of the gas and, therefore, work needs to be done on the system (i.e. gas) during the return journey. As a consequence, heat will be rejected by the system to the surroundings. This means that in a closed cycle (i.e. reversible) operation for the conversion of heat into work, a part of heat has to be rejected to the surroundings which should be at a temperature lower than that of the heat supplying system/substance. This results in change in thermodynamic states of the surroundings. The change in state of a system during the conversion of heat into work by the transfer of heat to the surroundings in closed cycle operation is called compensation. Thus, it can be stated that heat cannot be converted into work without compensation. Although this analysis has been made for an ideal gas under isothermal condition, the conclusions derived are of general validity and applicable to all processes.

When two bodies at different temperatures are brought in contact with each other, heat flows from the body at higher temperature to one at the lower temperature, irrespective of their physical properties and heat contents. In order to transfer the heat from a cold body to the hot one, one will have to employ a mechanical device in between the two. This device can receive heat from a cold body, raises its temperature and then deliver it to the hot body. All this requires work to be done by the device and only then heat can be transferred from a cold body to a hot body, i.e. without the additional work done such a heat transfer will not be possible. These and similar observations let to the statements which are known as the second law of thermodynamics.

3.3 Statements of Second Law of Thermodynamics

This law was first enunciated by Sadi Carnot in 1824 while doing theoretical studies on the efficiency of heat engines used for converting heat into work. The first Carnot's theorem states that the efficiency of machine operating with the Carnot cycle (i.e. a reversible cycle) depends only on the temperatures of the heater and cooler, and does not depend on the working substance used. In other words, all Carnot machines (i.e. reversible cycles) operating in the same temperature range have the same efficiency, irrespective of the working medium employed. The second Carnot's theorem states that the efficiency of any machine cannot be more than that of the Carnot machine working in the same temperature range, i.e. a reversible cycle has the highest efficiency. These concepts played an important role in the development of subsequent statements. In fact, Carnot's statement hardly makes any direct use of the energy concept. This is likely to be due to the fact that this statement was made much before the formulation of first law of thermodynamics. Therefore, it failed to be a general statement of second law and is simply referred to as Carnot's theorem.

The general statements of the second law, which include the energy concept, are due to Lord Kelvin and Clausius. Kelvin's statement can be stated as "It is just not possible to convert all heat into mechanical work without compensation". However, work can be completely converted to heat without compensation. It means that the entire absorbed heat is not converted into useful work and a part of it is rejected to the surrounding second reservoir. The above statement of Kelvin is concerned with the change in form of energy from heat to work, whereas the statement made by Clausius deals with the transfer of heat from one reservoir to another. This later version of the second law by Clausius states that it is impossible to transfer heat from a colder to a hotter body without producing permanent changes or compensation in the surroundings. This statement is based upon the performance of refrigerator—a heat engine working in the backward direction. A refrigerator transfers heat from a cold body to a hot body with the aid of an outside agency. Max Planck reformulated both the above postulates and suggested that it is impossible to construct a cyclic device/engine which can extract heat from a reservoir and produce no thermal effects elsewhere.

All the above statements are inductive in nature, i.e. they have been formulated by the generalization of experimental facts. The mathematical formulation of the aforesaid inductive statements led to the development of a new state function "Entropy" and the analytical statement of the second law was expressed by the following equation:

$$\mathrm{d}S \geq \frac{\delta q}{T} \tag{3.1}$$

where δq—an infinitesimal amount of heat supplied to the system, $\mathrm{d}S$—an infinitesimal increase in entropy of the system, and T—temperature of the surrounding which supplies the heat.

The above analytical statement of second law of thermodynamics embodied Kelvin, Clausius and Planck statements. Equality sign in the above equation corresponds to reversible processes (i.e. $\mathrm{d}S_{\text{rev.}} = \frac{\delta q}{T}$) and the inequality sign corresponds to irreversible processes (i.e. $\mathrm{d}S_{\text{irrev.}} > \frac{\delta q}{T}$).

3.4 Mathematical Formulation of Second Law of Thermodynamics

Mathematical formulation of this law, in general, starts with Kelvin's statement which states that heat cannot be converted into work without compensation. This means that in a closed cycle operation, a part of heat has to be rejected to the cold reservoir. In closed cycle operation, there is no change in internal energy of the system (i.e. $\Delta U = 0$), and it acts as a medium for the conversion of heat into work without undergoing any change in itself. Further, these cycles can operate either on reversible or irreversible conditions or a combination of both. As stated in Sect. 1.5.1, more work is done by the system in a reversible condition than that in an irreversible condition. Hence, in order to extract maximum work from the heat supplied to the system under a given set of conditions, cycles involving only reversible processes need to be used. But, the analytical analysis of such cases will lead to the mathematical formulation of second law for reversible processes only. However, it does not mean that one will have to go for another analysis for irreversible processes. In fact, the formulation for irreversible processes will follow from that for reversible processes. Hence, we shall confine ourself to mathematical formulation of this law for the reversible processes.

3.4.1 Conversion of Heat to Work by a Reversible Cycle

A reversible cycle consists of two or more reversible processes. For the conversion heat into work, the first process of this cycle needs to be reversible isothermal expansion in which the system, consisting of an ideal gas (say), will receive heat from the surroundings at a constant temperature and does work through expansion. This simultaneously changes the volume and pressure, i.e. the state of the system. In this process, the entire amount of heat absorbed (δq) by the system is directly converted into work (δw) with a simultaneous change in the state of the system. As per discussion made in Sect. 2.2.2, $\delta q = \delta w$ for this process, i.e. complete conversion of heat into work. But, one should not mean here that the second law has been violated by this complete conversion. In fact, the compensation in the present case is provided by the change in state of the system. Now, the system can be brought to the initial state by compression.

In order to achieve the initial state, some amount of heat must be extracted from the system under reversible conditions. There are a number of paths involving adiabatic, isothermal and polytropic processes by which it can be achieved. The simplest of these is a combination of adiabatic and isothermal processes as in this case, the exchange of heat with the surroundings under reversible condition can be achieved very easily because of the isothermal condition. Therefore, the whole cycle consists of four reversible processes as: (i) isothermal expansion of gas under reversible condition to get heat from the hot source, (ii) reversible adiabatic expansion of gas to bring down its temperature, (iii) reversible isothermal compression to reject a part of heat received by the system (i.e. an ideal gas) to a cold sink/reservoir, and (iv) reversible adiabatic compression to bring the system to its initial state. This cycle was first suggested by Sadi Carnot in 1824 and hence known as Carnot cycle. This cycle will now be analysed quantitatively for the extent of conversion of heat into work.

3.4.2 *Carnot Cycle with Ideal Gas as Working Substance*

As mentioned in Sect. 3.3, Carnot cycle, an ideal heat engine operating reversibly between a hot source at constant temperature (say T_2) and heat sink at other constant temperature (say T_1), consists of two isothermal (i.e. constant temperature) and two adiabatic (i.e. change in volume and pressure without loss or gain of heat) reversible processes/stages. These processes have been shown in their proper sequence in Fig. 3.1. The present discussion on Carnot cycle has been made in reference to an ideal gas as a working substance in the considered heat engine, but the principle involved is applicable to any thermodynamic substance. As indicated in this figure, the temperature of isothermal stage $1 \rightarrow 2$ is T_2 and that of $3 \rightarrow 4$ is T_1, and $T_2 > T_1$.

Isothermal expansion stage 1 → 2: In this stage, the gas receives q_2 amount of heat from the hot source at temperature T_2 and expands isothermally to state P_2, V_2 and performs work ($W_{1\rightarrow2}$). As discussed above, $\Delta U = 0$ and $q = w$ for an isothermal expansion/contraction of an ideal gas. Hence, we have the following:

$$q_{1\rightarrow2} = q_2 = W_{1\rightarrow2} = n\mathrm{RT}_2 \ln\left(\frac{V_2}{V_1}\right) \tag{3.2}$$

where n is the number of moles of gas.

Adiabatic expansion stage 2 → 3: The gas cylinder is removed from the hot reservoir, insulated and the gas is allowed to expand adiabatically to the state P_3, V_3, T_1. Here, $q_{2\rightarrow3} = 0$. First law for adiabatic process states that $w = -\Delta U$. Hence, we have,

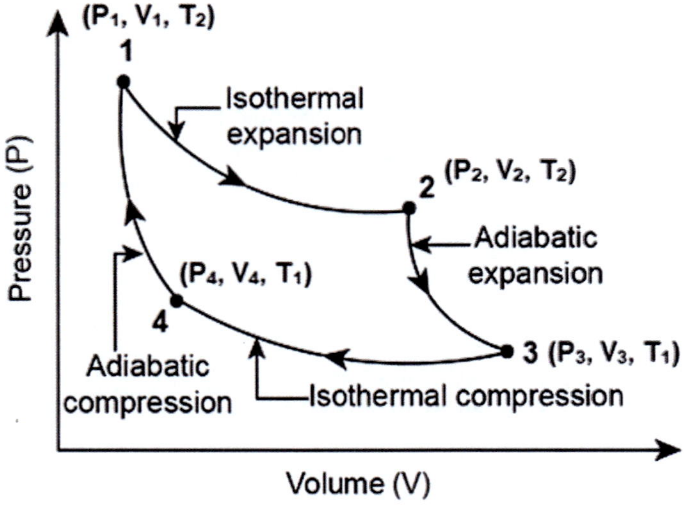

Fig. 3.1 Schematic $P - V$ diagram for Carnot cycle

$$W_{2\to 3} = -(\Delta U)_{2\to 3} = -n\int_{T_2}^{T_1} C_V dT = -nC_V(T_1 - T_2) \tag{3.3}$$

Isothermal compression stage 3 → 4: In this stage, the insulation is removed and the gas cylinder is connected to a cold reservoir (i.e. sink) at temperature T_1. Now, the gas is compressed isothermally to the state P_4, V_4, and it releases q_1 amount of heat to the sink. The work done on the system and heat released to the sink are given by the following equation:

$$q_1 = W_{3\to 4} = n\mathrm{RT}_1 \ln\left(\frac{V_4}{V_3}\right) \tag{3.4}$$

Adiabatic compression stage 4 → 1: In this stage, the cylinder is disconnected from the cold reservoir and surrounded by a perfect insulator so that no heat is exchanged with the surroundings. Now, the gas is compressed adiabatically to the initial state P_1, V_1, T_2. In this case, we have the following:

$$W_{4\to 1} = -(\Delta U)_{4\to 1} = -n\int_{T_1}^{T_2} C_V \mathrm{d}T = nC_V(T_1 - T_2) \tag{3.5}$$

Now, total work done (W_T) in the cycle = algebraic sum of work done in each of the processes, i.e. from Eqs. (3.2) to (3.5).

$$\begin{aligned} W_T &= W_{1\to 2} + W_{2\to 3} + W_{3\to 4} + W_{4\to 1} \\ &= n\mathrm{RT}_2 \ln\left(\frac{V_2}{V_1}\right) - nC_V(T_1 - T_2) \\ &\quad + n\mathrm{RT}_1 \ln\left(\frac{V_4}{V_3}\right) + nC_V(T_1 - T_2) \end{aligned}$$

or

$$\begin{aligned} W_T &= n\mathrm{RT}_2 \ln\left(\frac{V_2}{V_1}\right) + n\mathrm{RT}_1 \ln\left(\frac{V_4}{V_3}\right) \\ &= nR\left[T_2 \ln\left(\frac{V_2}{V_1}\right) + T_1 \ln\left(\frac{V_4}{V_3}\right)\right] = q_2 + q_1 \end{aligned} \tag{3.6}$$

We know that in an adiabatic process, $\Delta U = -w$. Hence, $C_V \mathrm{d}T = -\frac{\mathrm{RT}}{V}\mathrm{d}V$,

or

$$\frac{C_V}{R}\mathrm{d}(\ln T) = -\mathrm{d}(\ln V)$$

Integration of this equation between the limits V_4, T_1 and V_1, T_2 gives the following:

$$\ln\left(\frac{V_4}{V_1}\right) = \frac{C_V}{R}\ln\left(\frac{T_2}{T_1}\right) \tag{3.7}$$

Similarly,

$$\ln\left(\frac{V_3}{V_2}\right) = \frac{C_V}{R}\ln\left(\frac{T_2}{T_1}\right) \tag{3.8}$$

On equating Eqs. (3.7) and (3.8), we have.

$$\frac{V_4}{V_1} = \frac{V_3}{V_2} \text{ or } \frac{V_4}{V_3} = \frac{V_1}{V_2}.$$

Hence, Eq. (3.6) becomes as follows:

$$W_T = nR\left[T_2 \ln\left(\frac{V_2}{V_1}\right) - T_1 \ln\left(\frac{V_2}{V_1}\right)\right] = nR\ln\left(\frac{V_2}{V_1}\right)(T_2 - T_1)$$

or

$$W_T = nR\ln\left(\frac{V_2}{V_1}\right)(T_2 - T_1) \tag{3.9}$$

The efficiency of the reversible cycle or heat engine is defined as the ratio of work done (i.e. heat converted into work) to the heat absorbed by the system and is given by the following expression:

$$\eta = \frac{W_T}{q_2} = \frac{nR\ln\left(\frac{V_2}{V_1}\right)(T_2 - T_1)}{nRT_2\ln\left(\frac{V_2}{V_1}\right)} = \frac{(T_2 - T_1)}{T_2}$$

or

$$\eta = \frac{(T_2 - T_1)}{T_2} \tag{3.10}$$

Equation (3.10) shows that the efficiency of reversible cycle depends solely on the temperatures T_1 and T_2, i.e. the heat engine efficiency is independent of the working substance and depends only on the temperatures T_1 and T_2. T_1 cannot be zero and hence, complete conversion of heat into work is just not possible. In other words, all the heat supplied cannot be fully converted into work.

3.4.3 *Generalized Carnot Cycle*

In a generalized Carnot cycle, the working substance may be any thermodynamic material. The extension of Carnot cycle to any thermodynamic substance can be examined by consideration of a coupled system in which an ideal gas engine drives a second engine containing other substance through one complete reversible cycle, i.e. in forward and then in reverse direction, between the same temperature limits.

In this hypothetical system, engine, working with an ideal gas, transfers heat from a heat reservoir at temperature T_2 (say) to that at temperature T_1 (say), while an engine, working with another substance, transfers heat in the reverse direction, i.e. from heat reservoir at temperature T_1 to that at temperature T_2. The net work done by the engines in a complete cycle was theoretically proved to be zero, indicating that the efficiencies of both engines are the same. Hence, the generalized Carnot theorem is "all reversible heat engines, operating between the same temperature limits, must have the same efficiency".

3.5 Entropy

In going from one state to other, heat energy is either absorbed or liberated. The amount of heat absorbed or liberated is not constant but depends upon the temperature. Higher the temperature, more is the amount of heat absorbed or liberated and vice versa. If δq is the amount of heat absorbed or liberated at a temperature T in going from one state to other, then

$$\frac{\delta q}{T} = \text{constant}$$

This constant is called the change in entropy (i.e. dS) and thus the above equation can be written as:

$$\mathrm{d}S = \frac{\delta q}{T} \tag{3.11}$$

The concept of entropy was first introduced by Clausius in 1854 while working on the formulation of second law of thermodynamics and Eq. (3.11) is the required mathematical formulation of this law. In general, the change in entropy in passing from state A to B is given by the following equation:

$$\Delta S = \int_{S_A}^{S_B} \mathrm{d}S = S_B - S_A = \int_A^B \frac{\delta q}{T}$$

This expression for ΔS depends upon the thermodynamic coordinates (viz. T, P, V, U, etc.) of a system at the initial and final states. Hence, entropy of a system is a function of the thermodynamic coordinates defining the state of a system and its change (i.e. ΔS) between the two states is equal to the integral of the quantity $\delta q / T$. Though, entropy is an important thermodynamic quantity and has been proved to be very useful in the study of behaviours of heat engines, but the above mathematical formulation does not give any physical picture of entropy. Based on statistical concept, entropy of a substance has been defined as a measure of the degree of randomness or disorderness prevailing among its molecules and the growth of entropy implies an increase in the extent of disorderness in the arrangement of atoms/molecules in the substance. Hence, a high entropy system is in great disordered state or chaos.

Clausius summed up the second law by saying that the entropy of the universe tends to be a maximum, i.e. law of increase of entropy. Therefore, the general statement of the second law in terms of entropy may be enunciated as: "Every physical or chemical process in nature takes place in such a way so as to increase the entropy of the system." In order to examine the possibility of multiple processes explicitly, the consideration of the following mathematical form of second law becomes essential:

$$\Delta S_{\text{Universe}} = \Delta S_{\text{System}} + \Delta S_{\text{Surrounding}} = \sum \Delta S_i \geq 0$$

where each ΔS_i represents the entropy change for an individual process. Equation (3.11) provides a way to determine whether a process is reversible or irreversible. When the net entropy change is zero, the process is reversible and the system is said to be in equilibrium. If the total entropy is more than zero, the process will occur and such a process is often referred to as spontaneous or natural (i.e. irreversible) process. If the net entropy change is negative, the process is unnatural and will not occur. Thus, the second law of thermodynamics provides an important method for predicting the spontaneity or reversibility of a process.

3.5.1 Entropy: A State Property

Let us consider a Carnot cycle, i.e. a complete reversible process, as shown in Fig. 3.1. The entropy change associated with each of the four processes of a Carnot cycle can be calculated as follows:

Step I (i.e. in an isothermal expansion from state 1 to 2): In this stage, the working substance absorbs q_2 amount of heat energy from the hot reservoir at a constant temperature T_2. As heat is absorbed by the system, entropy change (ΔS_1) for this stage is positive and is calculated by using the following formula:

$$\Delta S_I = \int \frac{\delta q_{\text{rev}}}{T_2} = \frac{q_2}{T_2}$$

ΔS_I also represents gain in entropy of the working substance during its isothermal expansion.

Step II (i.e. an adiabatic expansion from state 2 to 3): During an adiabatic expansion, heat is neither given to nor removed from the system (i.e. $\delta q = 0$). Hence, $\Delta S_{II} = 0$.

Step III (i.e. an isothermal compression from state 3 to 4): In this process, the working substance gives out q_1 amount of heat energy to the sink (i.e. a cold reservoir) at a constant temperature T_1. There is a decrease in entropy of the working substance during this process due to the loss of heat energy and the change in entropy is calculated as:

$$\Delta S_{III} = \frac{q_1}{T_1}$$

Step IV (i.e. an adiabatic compression from state 4 to 1): During an adiabatic compression, no heat energy is either given to or removed from the system (i.e. $\delta q = 0$) and hence, there is no change in entropy during this process, i.e. $\Delta S_{IV} = 0$.

Thus, the entropy change for the complete cycle becomes as follows:

$$\Delta S_{\text{Cycle}} = \oint \mathrm{d}S = \Delta S_I + \Delta S_{II} + \Delta S_{III} + \Delta S_{IV} = \frac{q_2}{T_2} + \frac{q_1}{T_1}$$

In a complete reversible cycle, $\frac{q_2}{T_2} = -\frac{q_1}{T_1}$ or $\frac{q_2}{T_2} + \frac{q_1}{T_1} = 0\frac{q_2}{T_2} = -\frac{q_1}{T_1}$ or $\frac{q_2}{T_2} + \frac{q_1}{T_1} = 0$
Hence,

$$\Delta S_{\text{Cycle}} = \oint \mathrm{d}S = 0$$

A quantity, whose cyclic integral is zero, is a state property. Hence, the entropy is state function and depends only on the initial and final states of the system and not on the process path. The above equation means that the total change in entropy of the working substance in a complete cycle of reversible process is zero. Similarly, ΔS value for a combined system of heat source and sink is also zero. The cyclic summation of $q_{\text{rev}}/T = 0$ holds true for other reversible cycles also having any working substance either in solid, liquid or gaseous state.

3.6 Entropy Changes in Reversible Processes

During reversible processes, the system is never away from equilibrium. Such a process is, of course, imaginary. But, it is possible to conduct an actual process in a virtually reversible manner. Such an actual process is possible only under the influence of infinitesimally small driving force such that the system is very close to equilibrium. Whenever a system absorbs δq amount of heat, the surrounding loses an equal quantity of heat, i.e. $\delta q_{\text{System}} = -\delta q_{\text{Surrounding}}$, and it follows that $\frac{\delta q_{\text{system}}}{T} + \frac{\delta q_{\text{surrounding}}}{T} = 0$.

Consequently, entropy change ($\mathrm{d}S$ or ΔS) for a reversible process becomes zero. The following specific examples under this category are being presented here for discussion.

3.6.1 Reversible Isothermal Expansion of an Ideal Gas

We know that $\mathrm{d}S = \frac{\delta q_{\text{Rev}}}{T}$. Integration of this equation gives the following:

$$\int \mathrm{d}S = \int \frac{\delta q_{\text{Rev}}}{T}$$

Or

$$\Delta S_{\text{System}} = \frac{q_{\text{Rev}}}{T} \text{(since temperature is constant)} \tag{3.12}$$

As discussed in Sect. 2.2.2, $\mathrm{d}U = 0$ under isothermal condition. Hence, from the first law of thermodynamics, we have, $\delta q = P\mathrm{d}V = n\mathrm{RT}\frac{\mathrm{d}V}{V}$ (since, $PV = \mathrm{nRT}$ for n moles of an ideal gas). Integration of this equation yields the following:

$$q_{\text{Rev}} = n\mathrm{RT}\ln\left(\frac{V_2}{V_1}\right) \tag{3.13}$$

where V_1 and V_2 are the initial and final volumes of an ideal gas. Now, Eq. (3.12) yields the following:

$$\Delta S_{\text{System}} = \frac{q_{\text{Rev}}}{T} = nR\ \ln\left(\frac{V_2}{V_1}\right)$$

Or

$$\Delta S_{\text{System}} = nR\ \ln\left(\frac{V_2}{V_1}\right) \tag{3.14}$$

Since $V_2 > V_1$, $\Delta S_{\text{System}} > 0$. Expansion of gas causes the molecules to occupy a larger volume leading to an increase in entropy/randomness of the system.

As stated above, $\delta q_{\text{System}} = -\delta q_{\text{Surrounding}}$ or $q_{\text{System}} = -q_{\text{Surrounding}}$. Therefore,

$$\Delta S_{\text{Surrounding}} = -\frac{q_{\text{Rev}}}{T} = -nR\ln\left(\frac{V_2}{V_1}\right) \tag{3.15}$$

and $\Delta S_{\text{Total}} = \Delta S_{\text{System}} + \Delta S_{\text{Surrounding}} = 0$, as required for a reversible process.

3.6.2 Reversible Adiabatic Expansion of an Ideal Gas

All the adiabatic processes are accompanied without any exchange of heat energy between the system and the surrounding, i.e. $\delta q = 0$, and thus, $\Delta S_{\text{System}} = 0$, $\Delta S_{\text{Surrounding}} = 0$ and $\Delta S_{\text{Total}} = 0$. Hence, all the reversible adiabatic expansions and contractions are called isentropic processes because entropy remains unchanged.

3.6.3 Reversible Heating at Constant Volume or Variation of Entropy with Temperature Under Isochoric Condition

Second law gives,

$$\mathrm{d}S = \frac{\delta q}{T} \text{ or } \delta q = T\mathrm{d}S$$

From first law,

$$\mathrm{d}U = \delta q - P\mathrm{d}V = T\mathrm{d}S - P\mathrm{d}V$$

or

$$T\mathrm{d}S = \mathrm{d}U + P\mathrm{d}V$$

or

$$\mathrm{d}S = \frac{\mathrm{d}U}{T} + \frac{P}{T}\mathrm{d}V \tag{3.16}$$

Under isochoric condition, $\mathrm{d}V = 0$. Hence, Eq. (3.16) becomes $\mathrm{d}S = \frac{\mathrm{d}U}{T}$. As outlined in Sect. 2.3, we know that $\mathrm{d}U = C_V\mathrm{d}T$. This leads to the following:

$$\mathrm{d}S = C_V\frac{\mathrm{d}T}{T} \tag{3.17}$$

Heat capacity at constant volume (i.e. C_V) can be taken as constant over the considered temperature range of integration. This holds well if the temperature interval is small enough.

Let the temperature of the system changes from T_1 to T_2 and the corresponding change in entropy is from S_1 to S_2. Now, on integration of Eq. (3.17) between these limits, we have the following:

$$\int_{S_1}^{S_2} \mathrm{d}S = C_V \int_{T_1}^{T_2} \mathrm{d}\ln T$$

or

$$S_2 - S_1 = \Delta S = C_V \int_{T_1}^{T_2} \mathrm{d}\ln T \tag{3.18}$$

or

$$S_2 = S_1 + C_V \ln\left(\frac{T_2}{T_1}\right) \tag{3.19}$$

It is evident from Eq. (3.18) that if $T_2 > T_1$ then, $\Delta S > 0$. Heating of solid increases the energy of vibration of its molecules, and this results in an increase in disorderness. Equation (3.19) indicates that the value of entropy of a substance/system at a particular temperature T_2 can be calculated if the absolute value of the same at other temperature T_1 is known. But, the estimation of the absolute value of 'S' at a particular temperature is not possible from second law of thermodynamics. This problem was solved by the application of third law of thermodynamics, as discussed in Sect. (4.12).

The reversible flow of heat requires that the temperature of the system should differ infinitesimally (i.e. by a very small amount) from that of the surrounding. When the system is heated reversibly, $\Delta S_{\text{Surrounding}} = -\Delta S_{\text{System}}$ and thus $\Delta S_{\text{Total}} = 0$.

3.6.4 Reversible Heating at Constant Pressure or Variation of Entropy with Temperature Under Isobaric Condition

The change in entropy with temperature at constant pressure is analogous to that at constant volume except that C_P replaces C_V. We know that $\mathrm{d}S = \frac{\delta q}{T}$ or $\delta q = T\mathrm{d}S$ and $\mathrm{d}U = \delta q - P\mathrm{d}V = T\mathrm{d}S - P\mathrm{d}V$ or $T\mathrm{d}S = \mathrm{d}U + P\mathrm{d}V = \mathrm{d}\,(U + PV) = \mathrm{d}H$, since pressure is constant or

$$\mathrm{d}S = \frac{\mathrm{d}H}{T}$$

As outlined in Sect. 2.3, $\mathrm{d}H = C_P\mathrm{d}T$. This leads to the following:

$$\mathrm{d}S = \frac{C_P\mathrm{d}T}{T} = C_P\mathrm{d}\ln T \tag{3.20}$$

Let the temperature of the system changes from T_1 to T_2 and the corresponding change in entropy is from S_1 to S_2. Now, on integration of Eq. (3.20) between these limits and considering C_P being constant over the considered temperature range, we have the following:

$$\int_{S_1}^{S_2} \mathrm{d}S = C_P \int_{T_1}^{T_2} \mathrm{d}\ln T,$$

or

$$S_2 - S_1 = \Delta S = C_P \int_{T_1}^{T_2} \mathrm{d}\ln T \tag{3.21}$$

or

$$S_2 = S_1 + C_P \ln\left(\frac{T_2}{T_1}\right) \tag{3.22}$$

The discussions for these results are exactly the same as given in Sect. 3.6.3 for constant volume heating.

3.6.5 *Reversible Phase Transformation*

Any change from one phase to another, which may be either solid–solid phase transformation or melting of solid, boiling of liquid, sublimation of solid to gas and vice versa, is associated with an exchange of heat energy called enthalpy or latent heat of such transformation. These phase changes in single component systems occur at well-defined constant temperatures and pressures. For calculating ΔS values of the phase transformations, the heat must be transferred reversibly. Calculation of ΔS utilizes the following concepts:

(a) Heating up to the phase transformation temperature and cooling after the end of phase change to room temperature can be done reversibly by carrying them out very slowly. The entropy changes associated with these processes of heating and cooling can be calculated by using the following equations:

$$\text{At constant volume, } \mathrm{d}S = \frac{\delta q_{\text{Rev}}}{T} = \frac{\mathrm{d}U}{T} = \frac{C_V \mathrm{d}T}{T} = C_V \mathrm{d}\ln T.$$

$$\text{At constant pressure, } \mathrm{d}S = \frac{\delta q_{\text{Rev}}}{T} = \frac{\mathrm{d}H}{T} = \frac{C_P \mathrm{d}T}{T} = C_P \mathrm{d}\ln T.$$

(b) Phase transformations can be made to occur slowly and reversibly. For a pure substance (i.e. a single component system), reversible phase changes (e.g. melting, vapourization, sublimation, solid-state phase transformation, etc.) at a given pressure occurs isothermally. As discussed in Sect. 2.2.2, $q = \Delta H$ under an isobaric condition. Therefore, for melting of a pure substance,

$$\Delta S_m^0 = \frac{q_{\text{Rev}}}{T_m} = \frac{\Delta H_m^0}{T_m} = \frac{L_m}{T_m} \tag{3.23}$$

where ΔS_m^0 represents the entropy change of melting; ΔH_m^0 and L_m are the enthalpy change and latent heat of melting; and T_m is the melting point.

For boiling of a pure substance,

$$\Delta S_v^0 = \frac{q_{\text{Rev}}}{T_b} = \frac{\Delta H_v^0}{T_b} = \frac{L_v}{T_b} \tag{3.24}$$

where ΔS_v^0 represents the entropy change of vapourization; ΔH_v^0 and L_v are the enthalpy change and latent heat of vapourization; and T_b is the boiling point of the substance.

The general formula for any phase transformation in a pure substance is as follows:

$$\Delta S_{\text{tr}}^0 = \frac{q_{\text{Rev}}}{T_{\text{tr}}} = \frac{\Delta H_{\text{tr}}^0}{T_{\text{tr}}} \tag{3.25}$$

where ΔS_{tr}^0 represents the entropy change for phase transformation, and ΔH_{tr}^0 and T_{tr} are the enthalpy change and temperature for phase transformation. The above equation indicates that the change in entropy is equal to the reversible heat exchanged divided by the temperature at which the phase change occurs.

For an endothermic (i.e. $\Delta H = +$ ve) phase transformation, such as melting or vapourization, the heat is supplied to the system. Hence, entropy/disorderness of a liquid phase is higher than that of the corresponding solid phase, whereas that of the gaseous phase is the highest for any material. The entropy change in the surrounding during such isothermal phase transformation will be,

$$\Delta S_{\text{Surrounding}} = -\frac{\Delta H_{\text{tr}}^0}{T_{\text{tr}}}$$

(c) Let us considered an example in which a pure substance X undergoes the following process of phase transformation:

$$X \text{ (solid) at 298 K} \rightarrow X\text{(liquid) at } T_m \rightarrow X\text{(gas) at } T_b \rightarrow X\text{(gas) at } T$$

The entropy change associated with this process can be calculated by combining the above concepts as follows:

$$\Delta S = S_T^0 - S_{298}^0 = \int_{298}^{T_m} \frac{C_P^0(s)}{T} \mathrm{d}T + \Delta S_m^0 + \int_{T_m}^{T_b} \frac{C_P^0(l)}{T} \mathrm{d}T + \Delta S_v^0 + \int_{T_b}^{T} \frac{C_P^0(g)}{T} \mathrm{d}T \tag{3.26}$$

(d) For a solution, the value of entropy change depends on the composition of solution as well. Moreover, the equilibrium phase transformation temperature goes on changing as transformation progresses. Therefore, the simple equations like Eqs. (3.23), (3.24), (3.25) to (3.26) cannot be used. The entropy change in the formation of an ideal solution by mixing two or more components (e.g. A, B, C ...) has been discussed in Chap. 6 and the derived equation is as follows:

$$\Delta S_{\mathrm{mix}}^{\mathrm{ideal}} = -R(N_A \ln N_A + N_B \ln N_B + \cdots) \tag{3.27}$$

where, N_A, N_B,... are the mole fractions of components A, B, ... in the solution. Since, N_A, N_B,... < 1, ln N_A, ln N_B, ... < 0 and $\Delta S_{\mathrm{mix}}^{\mathrm{ideal}} > 0$. Hence, formation of solution is accompanied by an increase of entropy/disorderness.

3.7 Formulation of Second Law for Irreversible Processes

In Sect. 3.5, the treatment of second law has been confined only to reversible processes. In this section, the treatment of irreversible processes by second law has been discussed. Formulation of this law for irreversible processes can be arrived at by a comparison of the efficiency of a reversible cycle with that of an irreversible one. For this, a cyclic process has been considered in which heat is converted into work by an irreversible cycle 1 and the rejected heat to the cold reservoir is transferred to the hot reservoir by means of a reversible cycle 2, as shown in Fig. 3.2.

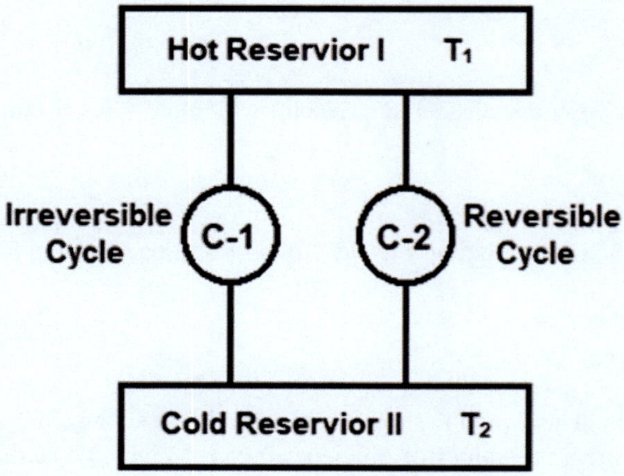

Fig. 3.2 Combined reversible and irreversible cycles for conversion of heat to work (C-1: Cycle-1 and C-2: Cycle-2)

Let q_1^I is the amount of heat taken by the irreversible cycle 1 from hot reservoir I at temperature T_1. A part of this absorbed heat is converted into work and the rest, say q_1^{II}, is rejected into the cold reservoir *II* at temperature T_2, where $T_1 > T_2$. The work done (i.e. w_1) by this irreversible cycle can be given by the expression,

$$w_1 = q_1^I - q_1^{II} \tag{3.28}$$

Now, consider the reversible cycle 2 which is to be used for transferring q_2^{II} amount of heat from cold reservoir *II* to the hot reservoir *I*. This can be achieved only by doing some mechanical work on the system. Let this work done be w_2(considered as negative as per convention), which can be written as follows:

$$w_2 = q_2^{II} - q_2^I \tag{3.29}$$

where q_2^{II} is the amount of heat taken by the reversible cycle 2 from cold reservoir *II* and q_2^I is the amount of heat supplied by this cycle to the hot reservoir *I*. The net work done by these cycles shall be,

$$w_1 + w_2 = \left(q_1^I - q_1^{II}\right) + \left(q_2^{II} - q_2^I\right) = \left(q_1^I - q_2^I\right) + \left(q_2^{II} - q_1^{II}\right)$$

or

$$w_1 + w_2 = q^I + q^{II} \tag{3.30}$$

where $q^I$and q^{II} are the net amounts of heat exchanged by the two reservoirs *I* and *II* through these cycles.

If η_{irr} and η_{rev} are the efficiencies of the irreversible cycle 1 and reversible cycle 2, then,

$$w_1 = \eta_{\text{irr}} \cdot q_1^I \tag{3.31}$$

$$w_2 = \eta_{\text{rev}} \cdot q_2^I \tag{3.32}$$

Let us considered a case for which $w_1 + w_2 = 0$ and in such a case,

$$w_1 + w_2 = \eta_{\text{irr}} \cdot q_1^I + \eta_{\text{rev}} \cdot q_2^I = 0 \tag{3.33}$$

On combining Eqs. (3.30) to (3.33), one gets the following:

$$q^I = -q^{II} = q_1^I\left(1 - \frac{\eta_{\text{irr}}}{\eta_{\text{rev}}}\right) \tag{3.34}$$

This equation shows that the net heat exchanged by each of the reservoir depends upon the values of efficiency, i.e. η_{irr} and η_{rev}. Here, there are three possibilities, viz. (i) $\eta_{\text{irr}} > \eta_{\text{rev}}$, (ii) $\eta_{\text{irr}} = \eta_{\text{rev}}$ and (iii) $\eta_{\text{irr}} < \eta_{\text{rev}}$. These possibilities are being discussed below in light of the statement of second law of thermodynamics to arrive at the feasible solution:

Case I: If $\eta_{\text{irr}} > \eta_{\text{rev}}$, then Eq. (3.34) shows that q^{I} is negative and q^{II} is positive. This means that the combined effect of two cycles will be to take heat from reservoir *II* (i.e. one at lower temperature) and deliver it to reservoir *I* (i.e. one at higher temperature) without any change in the working substance or compensation. This violates Clausius's statement of second law and hence is not feasible, i.e. η_{irr} cannot be more than η_{rev}.

Case II: If $\eta_{\text{irr}} = \eta_{\text{rev}}$, then it means that the work done in both the cycles is the same and they can be operated in reverse directions also. But, it has already been outlined in section that the work done in irreversible processes or cycles is less than that in reversible ones. Hence, $\eta_{\text{irr}} = \eta_{\text{rev}}$ is not feasible, i.e. η_{irr} cannot be equal to η_{rev}.

Case III: If $\eta_{\text{irr}} < \eta_{\text{rev}}$, then Eq. (3.34) shows that q^{I} is positive and q^{II} is negative. This implies that in this case, the combined effect of two cycles will be to absorb heat from reservoir *I* (i.e. one at higher temperature) and transfer it to reservoir *II* (i.e. one at lower temperature) without compensation. This is in conformity with the statement of the second law of thermodynamics.

From the above, the net conclusion is that the efficiency of the irreversible cycle is less than that of the reversible cycle and the statement $\eta_{\text{rev}} > \eta_{\text{irr}}$ leads to the formulation of second law for irreversible processes. If δq_{irr} and δq_{rev} are the amounts of heat converted into work in irreversible and reversible cycles, then their efficiencies η_{irr} and η_{rev} will be given by the following expressions:

$$\eta_{\text{irr}} = \frac{\delta q_{\text{irr}}}{q} \tag{3.35}$$

$$\eta_{\text{rev}} = \frac{\delta q_{\text{rev}}}{q} \tag{3.36}$$

Since, $\eta_{\text{rev}} > \eta_{\text{irr}}$, hence,

$$\delta q_{\text{rev}} > \delta q_{\text{irr}} \tag{3.37}$$

This equation forms the mathematical statement of the second law for irreversible processes and hold good not only for irreversible cycles but also for irreversible processes.

3.8 Spontaneous or Natural Processes

Each and every system tends to achieve a state that does not change with time and this is known as equilibrium state (i.e. a state of rest) of a system. A process which involves the spontaneous movement of a system from a non-equilibrium state to an equilibrium state is called a natural or spontaneous process. Such a process cannot be reversed without the application of an external agency, and hence, it is said to be an irreversible. The terms natural, spontaneous and irreversible are synonymous in this context. The flow of heat down a temperature gradient and mixing of gases are common examples of natural processes. When two bodies at different temperatures are placed in thermal contact with each other, the spontaneous flow of heat from a hotter to a colder body occurs and the equilibrium state is reached when both the bodies attain a common uniform temperature. In both these examples, the reverse process (i.e. the flow of heat up a temperature gradient un-mixing of gases) will never occur spontaneously. The simplicity of these systems along with common experience allows the equilibrium states to be predicted without any knowledge of the criteria for equilibrium. However, in less simple systems, the equilibrium states cannot be predicated from common experience and the criteria governing equilibrium must be established before making calculation for the equilibrium state.

Determination of equilibrium state is of prime importance in thermodynamics as knowledge of this state for any reaction system will give an idea whether the reaction will proceed in forward or backward direction, and in either case, to what extent before the establishment of equilibrium.

3.9 Entropy Change in Irreversible or Natural Processes

The Carnot cycle operates under reversible conditions. Let us suppose that the working substance in an engine performs an irreversible cycle of changes by absorbing q_1 amount of heat at a temperature T_1 from the hot source and rejecting q_2 amount of heat to the cold source (i.e. sink) at a temperature T_2. Then the efficiency of this cycle is given by the following expression:

$$\eta = \frac{q_1 - q_2}{q_1} = 1 - \frac{q_2}{q_1}$$

According to Carnot's theorem, this efficiency under irreversible condition is less than that of a reversible engine working between the same temperatures T_1 and T_2 for which,

$$\eta = \frac{T_1 - T_2}{T_1} = 1 - \frac{T_2}{T_1}$$

Thus,

$$1 - \frac{q_2}{q_1} < 1 - \frac{T_2}{T_1}$$

or

$$\frac{q_2}{q_1} > \frac{T_2}{T_1} \text{ or } \frac{q_2}{T_2} > \frac{q_1}{T_1}$$

or

$$\frac{q_2}{T_2} - \frac{q_1}{T_1} > 0 \tag{3.38}$$

where the term q_1 / T_1 represents the loss in entropy of the hot source and q_2 / T_2 is the term for gain in entropy of the cold source. Therefore, the net change in entropy (i.e. ΔS) for the whole system will be as follows:

$$\Delta S = \frac{q_2}{T_2} - \frac{q_1}{T_1} \tag{3.39}$$

The comparison of Eqs. (3.38) and (3.39) clearly indicates that ΔS is greater than zero or positive. Hence, there is an increase in entropy of the system during an irreversible process.

One may make this point still more clear by taking another example of an irreversible/natural process like conduction or radiation of heat. Let us consider a system consisting of two bodies at temperatures T_1 and T_2, where $T_1 > T_2$. Heat always flows from a higher to a lower temperature, either by conduction or radiation, and it is a natural process. Let q is the amount of heat transmitted in this process, then,

$$\text{Decrease in entropy of the hot body} = \frac{q}{T_1}.$$

$$\text{Increase in entropy of the cold body} = \frac{q}{T_2}.$$

Hence, the net increase in entropy of the system (i.e. ΔS) $= \frac{q}{T_2} - \frac{q}{T_1}$, which is a positive quantity because $T_1 > T_2$. Hence, the conclusion is that a natural or irreversible process generates entropy. Another example of frictional force can also be cited. The kinetic energy of a moving body gets decreased in a natural process by frictional forces and heat is generated, indicating an increase in entropy. We may, therefore, generalize the result by saying that the entropy of a system increases in all the irreversible/natural processes, and this increase is thus a measure of the degree of irreversibility. This is known as the law or principle of increase of entropy.

3.10 Degradation of Energy and Increase of Entropy

Degradation means conversion of energy from more useful form to less useful form. This can be illustrated clearly by taking an example of a spontaneous or natural process. When a system undergoes a spontaneous process, it comes nearer and nearer to the equilibrium state with passage of time and its capacity for further spontaneous change decreases. Finally, once equilibrium is established, the capacity of the system for doing further work gets exhausted. In the initial non-equilibrium state, some energy of the system is available for doing some useful work, and when the equilibrium state is reached as a result of completion of the spontaneous process, no any energy of the system is available for doing useful work. Thus, as a result of spontaneous/natural/irreversible process, the system becomes degraded in that energy which was available for doing useful work, i.e. the mechanical energy gets converted into thermal/heat energy which is not available for external purposes. In comparison to reversible process, any irreversible process performs less work. The difference between the work done during a reversible process and that done during an irreversible process (i.e. $w_{\text{rev}} - w_{\text{irr}}$) is the mechanical energy which has been degraded to thermal energy as a result of irreversible nature of the process. This heat energy, which is unavailable for useful work, gets added in the universe. In this way, all energies existing in different forms in the universe are expected to be converted into heat energy in a distant future and may not be available for conversion into mechanical work on account of irreversibility, i.e. the available energy of the universe for doing work will tend towards zero. This stage of the universe will correspond to a state of maximum degradation and entropy, and there will be no temperature gradient between various bodies of the universe due to convection, etc. After attainment of this the state, no any heat engine will be able to work because of no possibility of heat flow due to the uniformity of temperature throughout the universe. This implies that although the total amount of energy of the universe is conserved, it will get transformed into a form which will not be available for work. Thus, the energy is running downhill, and the universe is marching towards a stage of heat-death.

The entropy in the system is created or produced (i.e. $\Delta S > 0$) as a result of occurrence of an irreversible/natural/spontaneous process, i.e. the entropy is generated as a result of degradation of mechanical work into heat. In other words, the entropy produced is a measure of the degradation of energy which occurs as a result of the process. The zero entropy change during a reversible process (i.e. $\Delta S_{\text{rev}} = 0$) is due to the fact that no degradation occurs during a reversible process. Since, a reversible process represents a limiting ideal case, all actual processes are, in general, irreversible in nature. It means that as cycle after cycle of irreversible operation is performed, the entropy of the system increases and tends to achieve a maximum value. This is the principle of increase of entropy and may be stated as "the entropy of a system either increases or remains constant if the processes undergone by it are irreversible or reversible". Analytically, it may be expressed as $\mathrm{d}S \geq 0$, where the equality sign refers to reversible or equilibrium processes and the inequality sign to irreversible processes. Therefore, the necessary and sufficient condition of equilibrium of a system is that its entropy should be maximum. After attaining a maximum value, entropy cannot increase further and dS, will be equal to zero (i.e. condition for equilibrium). Within an increase in entropy, disorderness in the molecules of the substance increases, i.e. growth of entropy implies a transition from ordered to disordered or from less ordered to more ordered state. When temperature of the system is increased, the amounts of entropy and disorderness in it increase and vice versa. At $T = 0$ K, the thermal motion/agitation of molecules completely disappears and hence the entropy and disorderness of the molecules of a substance tend to zero, i.e. the molecules are in perfect order (i.e. 3rd law of thermodynamics). Entropy of a substance is, therefore, said to be a measure of the degree of disorderness prevailing among its molecules. One may recognize this fundamental law to be an inherent tendency of nature to proceed from a more ordered to a less ordered state or from a less ordered to a more disordered state. In other words, the ultimate destiny of the universe is not ordered but chaos.

3.11 Entropy Change Associated with Chemical Reaction

By definition, all the ongoing chemical reactions are not at equilibrium, i.e. they are irreversible in nature and accompanied by change in entropy. The total change in entropy in a chemical reaction, which may take place under constant T, P or V, remains the same irrespective of the fact whether it occurs in multiple stages or a single stage in the formation of final products from the reactants. Further, entropy is an extensive state property, and its values are additive. Hence, the entropy change for a reaction, occurring isothermally, can be obtained by summation of entropies of the reactants and the products in similar manner to the method previously described for the determination of heat of reaction. Let us consider the following chemical reaction occurring isothermally at temperature T:

$$X(\text{pure}) + AB(\text{pure}) = XA(\text{pure}) + B(\text{pure})$$

For this reaction, the entropy change at temperature T is calculated by the following equation:

$$\Delta S_T^0 = \sum S^o_{\text{products}} - \sum S^0_{\text{reacants}} = S^0_{XA}(\text{at } T) + S^0_B(\text{at } T) - S^0_X(\text{at } T) - S^0_{AB}(\text{at } T)$$

where $\sum S^o_{\text{reactants}}$ and $\sum S^o_{\text{products}}$ are the sum of entropies of reactants and the products. In view of irreversible nature of the reaction, use of the formula $\Delta S_T^0 = \Delta H^0/T$ for calculation of entropy change will not be correct.

Let us consider that during the course of a chemical reaction, the temperature changes from T_1 to T_2. The change in the value of ΔS^0 with temperature can be calculated by using the following equation:

$$\Delta S^0_{T_2} - \Delta S^0_{T_1} = \int_{T_1}^{T_2} \frac{\Delta C_P}{T} \mathrm{d}T$$

$$\Delta S^0_{T_2} = \Delta S^0_{T_1} + \int_{T_1}^{T_2} \frac{\Delta C_P}{T} \mathrm{d}T \tag{3.40}$$

Thus, change in entropy of a reaction at a particular temperature can be calculated from its value at any other temperature by making use of Eq. (3.40) which is similar to Kirchhoff's law. In the above equation, ΔC_P represents the difference between the sums of molar heat capacities of the products and the reactants at constant pressure.

If a phase change occurs either in the reactants or products between the considered temperature range of study (say T_1 to T_2), due allowance for the same needs to be made, as described in Sect. 3.6.5, for calculation of ΔS for the reaction. Qualitative assessment for the change in entropy during a chemical reaction can be made by having the knowledge of volume change (ΔV) accompanying a reaction. If $\Delta V\left(\sum V_{\text{products}} - \sum V_{\text{reactants}}\right)$ during a chemical reaction is positive, the disorderness/entropy in the system increases and vice versa. Let us consider the following decomposition reaction:

$$CaCO_3(s) \underset{\text{at } T>800\,^\circ\text{C}}{\overset{\text{Heating}}{\longrightarrow}} CaO(s) + CO_2(g)$$

In comparison to the volume of gas, volumes of solids may be considered as negligible and hence, $\Delta V = +ve$. As a result, disorderness/entropy in the system increase.

3.12 Entropy Change in Isolated Systems

An isolated system neither exchanges matter nor energy with its surroundings, i.e. it has no interaction with outside environment. From thermodynamic point of view, an isolated system contains the surroundings within itself.

For any isolated system, $\Delta S \geq 0$ depending upon whether the process is reversible or irreversible (i.e. natural), indicating that the entropy of a system in isolation can never decrease. It means that the conditions of a system in isolation cannot be varied in such a way so as to decrease the entropy. As known to us, reversible processes are rare. Hence, the entropy law for an isolated system can be stated as "the entropy of any isolated system increases and approaches, more or less rapidly, to the equilibrium state of maximum entropy".

3.13 Processing of Materials in Solid-State and Entropy Change

Equations used for the calculation of entropy change during solid-state phase transformations /treatments have already been outlined in Sects. 3.1 and 3.4. In this section, a qualitative assessment for the change in entropies of the materials, processed under the following solid-state metallurgical treatments, has been discussed:

(a) Graphitization of carbon
(b) Mechanical working of materials
(c) Hardening of steel
(d) Association and dissociation

(a) **Graphitization of Carbon**: Artificial graphite is manufactured by heating petroleum coke at a temperature of around 2700 °C in absence of air. During this graphitization treatment, the structure of the material used undergoes a change from complete amorphous state to perfectly crystalline state (i.e. HCP structure) as follows:

$$\underset{\text{(amorphous)}}{\text{Petroleum Coke}} \xrightarrow{\text{Heating at 2700}\,^{\circ}\text{C}} \underset{\text{(HCP structure)}}{\text{Graphite}}$$

As a result of this solid-state transformation in structure, the randomness in the material gets drastically reduced and the graphite obtained each ordered (i.e. HCP structure). Thermodynamically, the graphitization operation is accompanied by a decrease in entropy of the material/system, i.e. $\Delta S_{\text{graphitization}} < 0$ or negative.

(b) **Mechanical Working of Materials**: Mechanical working (i.e. rolling, forging, extrusion) involves the application of high pressure at a particular temperature to reduce the thickness of steel slabs or other materials and is of two types as: (i) hot mechanical working and (ii) cold mechanical working. Any mechanical working, which is carried out at a temperature above re-crystallization temperature of the material, is termed as hot mechanical working while that carried out at temperature below re-crystallization temperature is called cold mechanical working. During mechanical working, atoms/molecules in the material get displaced from their normal sites and as a result, dislocations are generated within the lattice.

Due to the involvement of high temperature, majority of the dislocations produced in hot worked material are cancelled out. However, a small amount of dislocations or randomness still persists in the hot worked material which leads to the generation of entropy in small amount, i.e. $\Delta S_{\text{HMW}} > 0$ or positive. On the other hand, cold mechanical working is done at relatively lower temperature. Hence, majority of the atoms/molecules do not return back to their normal sites and remain in their displaced positions, i.e. the cold worked materials have relatively higher dislocation density or randomness. As a result, entropy of the cold worked material gets drastically increased, i.e. $\Delta S_{\text{CMW}} > 0$ or positive and $\Delta S_{\text{CMW}} > \Delta S_{\text{HMW}}$.

(c) **Hardening of Steel**: Hardening involves heating of steel up to the austenitic stage followed by water or oil quenching. As a result, internal stresses are developed inside the matrix and material gets hardened. Hence, hardening of steel is accompanied by an increase of entropy.

(d) **Association and Dissociation**: Association means joining of smaller molecules into bigger ones and is accompanied by a decrease of entropy. Reverse is true for dissociation.

Solved Problems

Problem 3.1 Find out the value of standard entropy change (ΔS^o) for the reaction Fe_2O_3 (s) + 3C (s) = 2Fe (s) + 3CO (g) at 298 K. Data given for the values of standard entropy for various components at 298 K are as follows:

S^o for Fe_2O_3 (s) = 89.45 J/K mol, S^o for C (s) = 206.91 J/K mol, S^o for Fe (s) = 136.39 J/K mol and S^o for CO (g) = 197.71 J/K mol.

Solution

We know that

$$\begin{aligned}\Delta S^o_{298} &= \Sigma S^o_{298(\text{products})} - \Sigma S^o_{298(\text{reactants})}\\ &= \left[2 \times S^o_{298}\text{Fe(s)} + 3 \times S^o_{298}\text{CO(g)}\right] - \left[S^o_{298}\text{Fe}_2\text{O}_3\text{(s)} + 3 \times S^o_{298}\text{C(s)}\right]\\ &= 272.78 + 593.13 - 89.45 - 620.73 = 155.73\ \text{J/K mol}\end{aligned}$$

Problem 3.2 Determine the value of ΔS^o for the reaction Cr_2O_3 (s) + 3C (s) = 2Cr (s) + 3CO (g) at 298 K. Data given for the values of standard entropy for various components at 298 K are as follows:

S^o Cr_2O_3 (s) = 19.42 cal/K mol, S^o for C (s) = 1.36 cal/K mol, S^o for Cr (s) = 5.68 cal/K mol and S^o for CO (g) = 47.3 cal/K mol.

Solution

We know that

$$\begin{aligned}\Delta S^o_{298} &= \sum S^o_{298(\text{products})} - \sum S^o_{298(\text{reactants})}\\ &= \left[2 \times S^o_{298}\text{Cr(s)} + 3 \times S^o_{298}\text{CO(g)}\right] - \left[S^o_{298}\text{Cr}_2\text{O}_3\text{(s)} + 3 \times S^o_{298}\text{C(s)}\right]\\ &= 11.36 + 141.9 - 19.42 - 4.08 = 129.76\ \text{cal/K mol}\\ &= 542.4\ \text{J/K mol}\end{aligned}$$

Problem 3.3 Pure Fe (s) is heated from room temperature up to its melting point (1535 °C). Calculate the value of entropy of liquid Fe at this temperature with the help of the following data:

S^o_{298} for Fe (s) = 6.51 cal/K mol, latent heat of melting (L_m) = 3.68 kcal/mol and C_P for Fe (s) = 6.03 cal/K mol.

Solution
As is known to us,

$$\int_{298}^{1808} \mathrm{d}S = S^o_{1808} - S^o_{298} = \int_{298}^{1808} C_P \frac{\mathrm{d}T}{T} + L_m/T_m.$$

or

$$S^o_{1808}\mathrm{Fe(l)} = S^o_{298}\mathrm{Fe(s)} + \int_{298}^{1808} C_P\mathrm{Fe(s)} \frac{\mathrm{d}T}{T} + L_m/T_m$$

or

$$\begin{aligned} S^o_{1808}\mathrm{Fe(l)} &= 6.51 + 6.03[\ln 1808 - \ln 298] + 3680/1808 \\ &= 19.42 \text{ cal/K mol} = 81.16 \text{ J/K mol}. \end{aligned}$$

Problem 3.4 Determine the values of (a) standard entropy change (ΔS^o) associated with the heating of solid Cu from a temperature of 300 K to 1346 K and (b) entropy of Cu at a temperature of 1346 K with the help of the following data:

S^o_{300} for Cu (s) = 33.44 J/K mol and C_P for Cu (s) = $22.62 + 6.27 \times 10^{-3}\, T$ J/K mol.

Solution

(a) As is known to us,

$$\int_{300}^{1346} \mathrm{d}S = S^o_{1346} - S^o_{300} = \int_{300}^{1346} C_P \frac{\mathrm{d}T}{T}$$

or

$$\begin{aligned} \Delta S^o &= \int_{300}^{1346} \left(22.62 + 6.27 \times 10^{-3} T\right) \frac{\mathrm{d}T}{T} \\ &= 22.62[\ln 1346 - \ln 300] + 6.27 \times 10^{-3}[1346 - 300] \\ &= 33.96 + 6.56 = 40.52 \text{ J/K mol}. \end{aligned}$$

(b) Now,

$$S^o_{1346}\text{Cu(s)} = S^o_{300}\text{Cu(s)} + \int_{300}^{1346} (22.62 + 6.27 \times 10^{-3} T)\frac{dT}{T}.$$
$$= 33.44 + 40.52 = 73.96 \text{ J/K mol}.$$

Problem 3.5 Pure Au(l) is at a temperature of 1400 K and it is cooled to a temperature of 300 K. Find out value of standard entropy change (ΔS^o) during this process. Data given are as follows:

Melting point of Au = 1336 K, latent heat of melting (L_m) = 12.36 kJ/mol, C_P for Au (s) = $23.68 + 5.19 \times 10^{-3}T$ J/K mol and C_P for Au (l) = 29.30 J/K mol.

Solution
We know that,

$$\Delta S^o = \int_{1400}^{1336} C_P\text{Au(l)}\frac{dT}{T} - L_m/T_m + \int_{1336}^{300} C_P\text{Au(s)}\frac{dT}{T}$$
$$= \int_{1400}^{1336} 29.30 \times \frac{dT}{T} - 12{,}360/1336 + \int_{1336}^{300} (23.68 + 5.19 \times 10^{-3}T)\frac{dT}{T}$$
$$= 29.30(\ln 1336 - \ln 1400) - 9.25 + 23.68(\ln 300 - \ln 1336)$$
$$+ 5.19 \times 10^{-3}(300 - 1336)$$
$$= -1.37 - 9.25 - 35.37 - 5.38 = -51.37 \text{ J/K mol}$$

Problem 3.6 1 g mole of solid pure Au is heated from 298 K to its melting point (1336 K). Find out the value of standard entropy change (ΔS^o) with the help of the following data:

Latent heat of melting (L_m) = 12.36 kJ/mol

$$C_P \text{ for Au(s)} = 23.68 + 5.19 \times 10^{-3} \text{ T J/K mol}$$

Solution
We know that,

$$\Delta S^o = \int_{298}^{1336} C_P\text{Au(s)}\frac{dT}{T} + L_m/T_m$$
$$= \int_{298}^{1336} (23.68 + 5.19 \times 10^{-3}T)\frac{dT}{T} + 12{,}360/1336$$

$$= 23.68(\ln 1336 - \ln 298) + 5.19 \times 10^{-3}(1336 - 298) + 9.25$$
$$= 35.53 + 5.39 + 9.25 = 50.17\ \text{J/K mol.}$$

Problem 3.7 1 g mole of liquid pure Au at 1336 K is cooled to a temperature of 1250 K. Determine the value of ΔS^o in the system and surrounding with the help of the following data:

Latent heat of fusion $(L_m) = 12.36$ kJ/mol

$$C_P \text{ for Au(s)} = 23.68 + 5.19 \times 10^{-3}\ \text{T J/K mol}$$

Solution

We know that,

$$\Delta S^o_{\text{Sys}} = -L_m/T_m + \int_{1336}^{1250} C_P\text{Au(s)}\frac{\text{d}T}{T}$$
$$= -12{,}360/1336 + \int_{1336}^{1250} (23.68 + 5.19 \times 10^{-3}T)\frac{\text{d}T}{T}$$
$$= -9.25 + 23.68(\ln 1250 - \ln 1336) + 5.19 \times 10^{-3}(1250 - 1336)$$
$$= -9.25 - 1.58 - 0.45 = -11.28\ \text{J/K mol.}$$

The system (liquid Au) releases heat and the surrounding absorbs heat during freezing of liquid Au up to 1250 K.

We know that, $\Delta S^o_{\text{Surr}} = -\frac{\Delta H^o_{\text{Sys}}}{T} = -\frac{\Delta H^o_{\text{Sys}}}{1250}$.

Now,

$$\Delta H^o_{\text{Sys}} = -L_m + \int_{1336}^{1250} C_P\text{Au(s)d}T$$
$$= -12{,}360 + \int_{1336}^{1250} (23.68 + 5.19 \times 10^{-3}T)\text{d}T$$
$$= -12{,}360 + 23.68(1250 - 1336) + 2.6 \times 10^{-3}(1250^2 - 1336^2)$$
$$= -12{,}360 - 2036.48 - 578.23 = -14{,}974.71\ \text{J/mol}$$

or $\Delta S^o_{\text{Surr}} = -\frac{\Delta H^o_{\text{Sys}}}{1250} = 11.98$ J/K mol

$$\Delta S^o_{\text{Universe}} = \Delta S^o_{\text{Sys}} + \Delta S^o_{\text{Surr}} = -11.28 + 11.98 = 0.7\ \text{J/K mol}$$

Problem 3.8 Value of entropy of pure Zn at 298 K is 9.96 cal/K mol, and it melts at 693 K. Find out the values of standard entropy change associated with the heating of pure Zn up to 1023 K and entropy of pure Zn at 1023 K. Data given are as follows:

Latent heat of melting (L_m) of Zn = 1.74 kcal/mol, C_P for Zn (s) = 5.35 + 2.40 × 10^{-3} T cal/K mol and C_P for Zn (l) = 7.51 cal/K mol.

Solution
As is known to us,

$$\int_{298}^{1023} \mathrm{d}S = S^o_{1023} - S^o_{298} = \Delta S^o$$

$$\begin{aligned}\Delta S^o &= \int_{298}^{693} C_P\mathrm{Zn(s)}\frac{\mathrm{d}T}{T} + \frac{L_m}{T_m} + \int_{693}^{1023} C_P\mathrm{Zn(l)}\frac{\mathrm{d}T}{T}\\ &= \int_{298}^{693} (5.35 + 2.40 \times 10^{-3}T)\frac{\mathrm{d}T}{T} + 1740/693 + \int_{693}^{1023} 7.51 \times \frac{\mathrm{d}T}{T}\\ &= 5.35(\ln 693 - \ln 298) + 2.40 \times 10^{-3}(693 - 298)\\ &\quad + 2.51 + 7.51(\ln 1023 - \ln 693)\\ &= 4.52 + 0.82 + 2.51 + 2.92 = 10.77\end{aligned}$$

or

$$S^o_{1023} = S^o_{298} + 1077 = 9.96 + 10.77 = 20.73\ \mathrm{cal/K\,mol} = 86.65\ \mathrm{J/K\,mol}$$

Problem 3.9 1000 g of pure solid Pb is heated from 300 to 550 K under constant pressure condition. Find out the value of standard entropy change (ΔS^o) during this heating process with the help of the following data.

$$C_P \text{ for Pb(s)} = 5.63 + 2.33 \times 10^{-3}\ \mathrm{T\ cal/K\,mol}.$$

Solution
We know that atomic weight of Pb = 208 g, no. of moles of Pb (n_{Pb}) = 1000/208 = 4.8

$$\begin{aligned}\Delta S^o &= n_Pb \times \int_{300}^{550} C_P\mathrm{Pb(s)}\frac{\mathrm{d}T}{T} = 4.8 \times \int_{300}^{550} (5.63 + 2.33 \times 10^{-3}T)\frac{\mathrm{d}T}{T}\\ &= 4.8\left[5.63(\ln 550 - \ln 300) + 2.33 \times 10^{-3}(550 - 300)\right]\end{aligned}$$

$$= 4.8[3.41 + 0.58] = 19.15\ \text{cal/K}$$

Problem 3.10 1 mol of oxygen gas having $C_V = 25.92$ J/K mol is heated from 300 to 700 K under isochoric condition. Find out the value of standard entropy change (ΔS^o) during this heating process.

Solution

$$\Delta S^o = \int_{300}^{700} C_V O_2(g)\frac{dT}{T} = \int_{300}^{700} 25.92\frac{dT}{T}$$
$$= 25.92(\ln 700 - \ln 300) = 21.96\ \text{J/K mol}$$
$$= 5.25\ \text{cal/K mol}$$

Problem 3.11 Melting point of Pb at a pressure of 1 atm is 600 K. At this temperature the latent heat of fusion of Pb is 4807 J/mol. After this, the liquid Pb spontaneously freezes to 590 K at 1 atm pressure. Data given are as follows:

$$C_P \text{ for Pb(s)} = 23.53 + 9.74 \times 10^{-3}\ \text{T J/K mol}$$

$$C_P \text{ for Pb(l)} = 32.40 - 3.10 \times 10^{-3}\ \text{T J/K mol}$$

Calculate values of the followings:

(a) ΔS and ΔH during heating from 590 to 600 K.
(b) ΔS and ΔH at the melting and freezing points.
(c) ΔS and ΔH during cooling from 600 to 590 K.
(d) Total entropy and enthalpy changes.

Solution

(a) As per definition, ΔS during heating from 590 to 600

$$K = \int_{590}^{600} C_P\text{Pb(l)}\frac{dT}{T} = \int_{590}^{600} (32.40 - 3.10 \times 10^{-3}T)\frac{dT}{T} = 32.40$$
$$\times \ln(600/590) - 3.10 \times 10^{-3}(600 - 590)$$
$$= 0.514\ \text{J/K mol}$$

ΔH during heating from 590 to 600

$$K = \int_{590}^{600} C_P\text{Pb(l)}\mathrm{d}T = \int_{590}^{600} \left(32.40 - 3.10 \times 10^{-3}T\right)\mathrm{d}T$$
$$= 32.40(600 - 590) - 1.55 \times 10^{-3}\left(600^2 - 590^2\right)$$
$$= 305.55\ \mathrm{J/mol}$$

(b) As per definition, ΔS at the melting point $= L_m/T_m = 4807/600 = -8.04\ \mathrm{J/K\,mol}$.

ΔS at the freezing point $= L_{\text{freez}}/T_{\text{freez}} = -4807/600 = -8.04\ \mathrm{J/K\,mol}$.
ΔH at the melting point = 4807 J/mol, ΔH at the freezing point = – 4807 J/mol,

(c) As per definition, ΔS during cooling from 600 to 590

$$K = \int_{600}^{590} C_P\text{Pb(s)}\mathrm{d}T = \int_{600}^{590} \left(23.53 + 9.74 \times 10^{-3}T\right)\frac{\mathrm{d}T}{T}$$
$$= 23.53 \times \ln(590/600) + 9.74 \times 10^{-3}(590 - 600)$$
$$= -0.4924\ \mathrm{J/K\,mol}$$

ΔH during freezing from 600 to 590

$$K = \int_{600}^{590} C_P\text{Pb(s)}\mathrm{d}T = \int_{590}^{600} \left(23.53 + 9.74 \times 10^{-3}T\right)\mathrm{d}T$$
$$= 23.53(590 - 600) + 4.87 \times 10^{-3}\left(590^2 - 600^2\right)$$
$$= -293.25\ \mathrm{J/mol}$$

(d) Total entropy change during freezing $(\Delta S_T) = 0.514 + (-8.04) + (-0.4924)\ \mathrm{J/K\,mol} = -8.02\ \mathrm{J/K\,mol}$.

Total enthalpy change during freezing $(\Delta H_T) = 305.55 + (-4807) + (-293.25) = -4794.7\ \mathrm{J/mol}$.

Exercise

3.1 Cu boils at a temperature of 2848 K and the latent heat of vapourization (enthalpy change) at this temperature is 72.82 kcal/mol. Find out the value of standard entropy change at its boiling point.

[**Ans**: 106.88 J/K mol].

3.2 Pure Fe melts at a temperature of 1808 K and 1 atm pressure. Its latent heat of melting (enthalpy change) at the melting point is 3.30 kcal/mol. Find out the value of standard entropy change at the melting point of Fe.

[**Ans**: 7.634 J/K mol].

3.3 Pure Cu is heated from a temperature of 25–1527 °C. Calculate the values of standard entropy change involved in this heating process and entropy of Cu at a temperature of 1527 °C with the help of the following data:

Melting point of Cu = 1083 °C, C_P for Cu (s) = 24.5 J/K mol (25–1527 °C), S^o_{298} Cu (s) = 33.3 J/K mol and latent heat of melting of Cu = 12.97 kJ/mol.

[**Ans**: 33.65 J/K mol; 86.95 J/K mol].

3.4 Pure Pb is heated from a temperature of 25–727 °C. Find out the values of standard entropy change during this heating process and entropy of Pb at 727 °C. Data given are as follows:

Melting point of Pb = 372 °C, C_P for Pb (s) = 23.6 + 9.75 × 10^{-3} T J/K mol, C_P for Pb (l) = 32.4 – 3.1 × 10^{-3} T J/K mol, S^o_{298} Pb (s) = 64.9 J/K mol and latent heat of melting of Pb = 4.81 kJ/mol.

[**Ans**: 42.79 J/K mol; 107.69 J/K mol].

3.5 Calculate the Values of ΔH^o and ΔS^o During Heating of Solid Aluminium (Al):

(i) From room temperature to 727 °C
(ii) From its melting point of 660 °C to a temperature of 727 °C

Data given are as follows:

Latent heat of fusion at the melting point = 2607.65 cal/mol
C_P for Al(s) = 7.78 cal/K mol and C_P for Al(l) = 7.01 cal/K mol

[**Ans**: (i) ΔH^o = 33513.65 J/mol; ΔS^o = 50.83 J/mol; (ii) ΔH^o = 12863.2 J/mol; ΔS^o = 13.71 J/K mol].

3.6 Outline the expression for $S_T - S_{298}$ in terms of temperature T, if $C_P = a + bT + cT^{-2}$.

3.7 Prove the following: $C_P - C_V = \left(\frac{\alpha^2 VT}{\beta}\right)$

$$\text{where } \alpha = \frac{1}{V}\left(\frac{\delta V}{\delta T}\right)_P \text{ and } \beta = -\frac{1}{V}\left(\frac{\delta V}{\delta P}\right)_T$$

3.8 The Value of C_P of RbF in the Temperature Range of 25 °C up to Its Melting Temperature of 775 °C is as Follows:

$$C_P \text{ for RbF(s)} = 33.31 + 38.46 \times 10^{-3}T + 5.06 \times 10^5 T^{-2} \text{ J/K mol}$$

In the temperature range of 775–927 °C, the C_P value of liquid RbF is given as follows:

$$C_P \text{ for RbF(l)} = -47.23 + 3.48 \times 10^{-3}T + 1465.93 \times 10^{5}T^{2} \text{ J/K mol}$$

Enthalpy change at the melting point of RbF = 26,334 J/mol.
Find out the value of ΔS during heating of RbF from 27 to 927 °C.

[**Ans**: 108.14 J/K mol].

Chapter 4
Free Energy, Criteria for Equilibrium and Thermodynamic Relations

This chapter, consisting of twelve sections, deals with the detailed illustration on potential topics which are of high importance in predicting the feasibility of reactions/ processes, stability of reactants and products, establishment of equilibrium in the system, etc. Various expressions for the combined statements of first and second laws of thermodynamics, and establishment of criteria for the attainment of equilibrium under different conditions have been outlined substantially in Sects. 4.2 and 4.3 of this chapter. A detailed discussion on Gibbs and Helmholtz free energies along with their differences, change in Gibbs free energy (ΔG) and feasibility of a reaction, and standard states have been introduced in Sects. 4.1, 4.4, and 4.5. Dependence of ΔG on temperature and pressure has also been elaborated and equations showing relations between them have been established in Sect. 4.6. A broad knowledge on the derivation of Gibbs–Helmholtz and Clausius–Clapeyron equations along with their applications in Metallurgy has been presented in Sects. 4.7 and 4.9.

In addition to the above, a thorough discussion on the latest statement of third law of thermodynamics along with its verification, consequences and importance have been covered in this chapter. This chapter also includes a short description of Richard's and Trouton's rules. At the end of this chapter, sufficient number of influential and critical problems have been solved and given as an exercise.

4.1 Free Energy

It is an extensive property and is proportional to mass of the substance. In thermodynamics, free energy is expressed as per g-mole of the substance. However, it also refers to the entire system. The free energy of a substance depends upon the following factors:

(a) Mass of the substance.
(b) Chemical composition of the substance.

S. K. Sahoo et al., *Fundamentals of Metallurgical Thermodynamics*,
https://doi.org/10.1007/978-981-99-6671-4_4

(c) Structure of the material (e.g. crystal structure, microstructure, macrostructure, etc. in case of solid).
(d) Strengths of other energy fields (e.g. gravitational, electric, magnetic, etc.).
(e) Specific surface area.

We know the followings:
$dU = \delta q - \delta w$ from 1st law of thermodynamics and
$\delta q_{rev} = TdS$ from 2nd law of thermodynamics.
Combination of these two equations yields,

$$dU = TdS - \delta w \text{ for a reversible process} \tag{4.1}$$

As already discussed in Chap. 3, efficiency of the reversible process is maximum, i.e. the reversible process performs maximum work. Hence, the term δw in Eq. (4.1) corresponds to maximum work. Thus,

$$\delta w(\text{maximum work}) = TdS - dU$$

This maximum work is the sum of mechanical work done against pressure (i.e. PdV) and non-mechanical work (e.g. chemical work—done in chemical reaction, electrical work—done in electrochemical cell, etc.). In a reversible process, the system performs no work other than that done against pressure. Hence, Eq. (4.1) can be written in the following form:

$$dU = TdS - PdV$$

In heat engines, mechanical work done is of main interest whereas in chemical processes, non-mechanical work done is of prime importance. In order to know the capability of the system to do work, one will have to find out its free energy, i.e. the energy available for work. The following two free energy functions, represented after the names of the scientists who first derived them, are employed in thermodynamics:

(i) Helmholtz Free Energy:

$$A = U - TS \tag{4.2}$$

(ii) Gibbs Free Energy:

$$G = H - TS \tag{4.3}$$

As is clear from Eqs. (4.2) and (4.3), Helmholtz free energy utilizes the concept of internal energy (U) and the Gibbs free energy utilizes the concept of enthalpy (H). A and G are state properties of the system because U, H, S and T are all state functions (i.e. U, H and S—properties and T—variable). Secondly, the values of A and G are additive because they are extensive properties. The unit of entropy is JK^{-1}. Hence, the quantity TS has a unit of energy (i.e. joule). Basically, TS represents bond energy which cannot be utilized for doing work, but gets dissipated as heat. Helmholtz and Gibbs free energies can be used to predict the feasibility of spontaneity of a chemical reaction under a given set of conditions imposed on the system, and also to determine the conditions under which the system is at equilibrium.

4.1.1 Helmholtz Free Energy

It is defined by the relation $A = U - TS$. For a process/reaction occurring at constant temperature, this equation can be written as follows:

$$\Delta A = \Delta U - T\Delta S \text{ for finite changes} \tag{4.4}$$

Or

$$\mathrm{d}A = \mathrm{d}U - T\mathrm{d}S \text{ for infinitesimal changes} \tag{4.5}$$

where ΔA, ΔU and ΔS are the finite changes in Helmholtz free energy, internal energy and entropy. According to second law of thermodynamics, the heat exchanged during a reversible process is given by the following equation:

$$q_{\text{rev}} = T\Delta S$$

With this relationship, Eq. (4.4) takes the following form:

$$q_{\text{rev}} = \Delta U - \Delta A \tag{4.6}$$

From the 1st law, q_{rev} and w_{rev} are related by the expression,

$$q_{\text{rev}} = \Delta U + w_{\text{rev}} \tag{4.7}$$

A comparison of Eq. (4.6) with Eq. (4.7) yields $w_{\text{rev}} = -\Delta A$ **or** in terms of infinitesimal changes, $\delta w_{\text{rev}} = -\mathrm{d}A$.

Here, w_{rev} or δw_{rev} represents the reversible maximum work done (i.e. mechanical work and non-mechanical work together) by the system. These expressions show that the change in Helmholtz free energy is equal and opposite in sign to the reversible work, i.e. the maximum amount of work done by the system is equal to decrease in the value of A. Therefore, Helmholtz free energy acts like a store of work or energy available for doing useful work by the system and is thus also called maximum work function.

4.1.2 *Gibbs Free Energy*

Helmholtz free energy is defined in terms of changes in temperature and volume during the reaction, but majority of the chemical and metallurgical processes in industries and laboratories are carried out more readily by controlling their temperatures and pressures. These conditions are met by using Gibbs free energy. Hence, it is the most important thermodynamic function from the point of view of applications. In chemical thermodynamics, we are mainly interested in non-mechanical work. Now,

$$\begin{aligned}\text{Reversible non-mechanical work} &= \delta w - P\mathrm{d}V = -(\mathrm{d}U - T\mathrm{d}S) - P\mathrm{d}V \\ &= -(\mathrm{d}U + P\mathrm{d}V - T\mathrm{d}S) = -(\mathrm{d}H - T\mathrm{d}S)\end{aligned} \tag{4.8}$$

(Since, $H = U + PV$ and $\mathrm{d}H = \mathrm{d}U + P\mathrm{d}V$ at constant pressure)
As mentioned above, Gibbs free energy is represented by the following relation:

$$G = H - TS$$

Differentiation of this equation at constant temperature and pressure yields,

$$\mathrm{d}G = \mathrm{d}H - T\mathrm{d}S \tag{4.9}$$

On comparing Eqs. (4.8) and (4.9), we get the following:

$$\text{Reversible non-mechanical work} = \delta w - P\mathrm{d}V = -\mathrm{d}G.$$

This expression states that the change in Gibbs free energy during a process is equal and opposite in sign to the reversible non-mechanical work, i.e. the non-mechanical work done by the system is equal to decrease in the value of *G*. Hence, Gibbs free energy acts as a store of non-mechanical work or energy available for doing non-mechanical work by the system.

4.2 Expressions for the Combined Statements of First and Second Laws of Thermodynamics

Assumptions: (i) a close system (i.e. fixed mass and composition) and (ii) a reversible process.

From the 1st law of thermodynamics, we have

$$\mathrm{d}U = \delta q - \delta w = \delta q - P\mathrm{d}V - \delta w' \tag{4.10}$$

2nd law of thermodynamics yields

$$\delta q = T\mathrm{d}S \tag{4.11}$$

On combining these equations, we have the following:

$$\mathrm{d}U = T\mathrm{d}S - P\mathrm{d}V - \delta w' \tag{4.12}$$

Again, by definition: $H = U + PV$
Differentiation of this equation gives the following:

$$\mathrm{d}H = \mathrm{d}U + P\mathrm{d}V + V\mathrm{d}P \tag{4.13}$$

On combining Eqs. (4.12) and (4.13), we have the following:

$$\mathrm{d}H = T\mathrm{d}S - P\mathrm{d}V - \delta w' + P\mathrm{d}V + V\mathrm{d}P = T\mathrm{d}S + V\mathrm{d}P - \delta w' \tag{4.14}$$

On differentiating Eq. (4.2), we get the following:

$$\mathrm{d}A = \mathrm{d}U - T\mathrm{d}S - S\mathrm{d}T \tag{4.15}$$

On combining Eqs. (4.12) and (4.15), we obtain the following:

$$\mathrm{d}A = T\mathrm{d}S - P\mathrm{d}V - \delta w' - T\mathrm{d}S - S\mathrm{d}T = -P\mathrm{d}V - S\mathrm{d}T - \delta w' \tag{4.16}$$

On differentiating Eq. (4.3), we get the following:

$$\mathrm{d}G = \mathrm{d}H - T\mathrm{d}S - S\mathrm{d}T \tag{4.17}$$

On combining Eqs. (4.14) and (4.17), we have the following:

$$\mathrm{d}G = T\mathrm{d}S + V\mathrm{d}P - \delta w' - T\mathrm{d}S - S\mathrm{d}T = V\mathrm{d}P - S\mathrm{d}T - \delta w' \tag{4.18}$$

In the above equations, $P\mathrm{d}V$ represents the mechanical work done against pressure and $\delta w'$ stands for non-mechanical work. Equations (4.12), (4.14), (4.16) and (4.18) have been derived by combining first and second laws of thermodynamics with the assumptions made above. If, a system is capable of doing only mechanical work (i.e. $\delta w'$), then these equations get further simplified into the followings:

$$\mathrm{d}U = T\mathrm{d}S - P\mathrm{d}V \tag{4.19}$$

$$\mathrm{d}H = T\mathrm{d}S + V\mathrm{d}P \tag{4.20}$$

$$\mathrm{d}A = -S\mathrm{d}T - P\mathrm{d}V \tag{4.21}$$

$$dG = -SdT + VdP \tag{4.22}$$

All these equations for combined statements include only those terms which are state functions and hence are exact differential equations. These relations are, therefore, independent of path.

4.3 Criteria of Equilibrium: Thermodynamic Potentials

There must be a driving force to move the system from one state to another, which is commonly referred to as thermodynamic potential. Heat flows from a body at higher temperature to that at a lower temperature when they are brought in contact. Thus, the temperature difference acts as a potential for heat flow. Similarly, the high pressure of a gas acts as a potential for mechanical work to be done by the gas. Also, the higher electrical voltage is the potential for the flow of current. From the above discussion, the following points emerge as the characteristics of potential:

(i) In a natural process, the system tends to move from a state of higher to lower potential.
(ii) For equilibrium to exist, the potentials of all the systems or sub-systems must be the same.

In this section, the thermodynamic potentials governing the attainment of equilibrium in chemical and metallurgical processes, exposed to different conditions, have been deduced and discussed with the help of Eqs. (4.19) to (4.22). In order to develop these criteria, Max Planck classified thermodynamic processes in three categories, viz. natural, equilibrium and unnatural. Both the natural and equilibrium processes have been defined in Chap. 3. Unnatural processes behave in a manner that is reverse of the natural processes. He further stated that the state of equilibrium of a system lies between the natural and unnatural processes.

4.3.1 Criteria of Equilibrium Under Constant Volume and Entropy Conditions

Mathematically, these conditions are expressed as $dV = 0$ and $dS = 0$. The constancy of entropy, i.e. isentropic condition, means $dS = 0$ or $\delta q / T = 0$. In general, T is not zero except the extreme case of absolute zero state. Therefore, δq will have to be zero. In other words, an isentropic process is adiabatic and no exchange of heat between the system and its surrounding is possible. The expression for the first combined statement of 1st and 2nd laws of thermodynamics, as stated in Eq. (4.19), is as follows:

$$dU = TdS - PdV$$

On inserting $dV = 0$ and $dS = 0$, it reduces to $dU = 0$ or $(\delta U)_{S,V} = 0$ (4.23)

Thus, internal energy becomes the necessary and sufficient criteria for the establishment of equilibrium under constant entropy and volume conditions.

As illustrated in Sect. 3.7, $\delta q_{irr} < \delta q_{rev}$ and by definition, $\delta q_{rev} = TdS$.

Hence, $\delta q < TdS$ or $\delta q - PdV < TdS - PdV$ for an irreversible process

Or $dU < TdS - PdV$ for an irreversible process

Under the specified conditions (i.e., $dV = 0$ and $dS = 0$),

$$dU < 0 \text{ for an irreversible or natural process} \tag{4.24}$$

and

$$\text{just the reverse, i.e. } dU > 0, \text{ for an unnatural process} \tag{4.25}$$

Therefore, the net result is $dU <=> 0$ for natural, equilibrium (i.e. reversible) and unnatural processes. Equations (4.24) and (4.25) indicate that under isochoric and isentropic conditions, the natural course of a process is in a direction of lower internal energy whereas for an unnatural process, it is in the direction of higher internal energy. Equation (4.23) represents the equilibrium state of a system. This discussion reveals that the internal energy fulfils the properties of a potential, as outlined in Sect. 4.3, and thus qualifies to be called a thermodynamic potential for the systems under constant volume and entropy conditions. Therefore, the internal energy is generally expressed as a function of entropy and volume as follows:

$$U = U(S, V)$$

One can write the following expression for the total differential of internal energy:

$$dU = \left(\frac{\delta U}{\delta S}\right)_V dS + \left(\frac{\delta U}{\delta V}\right)_S dV \tag{4.26}$$

A comparison of this expression with Eq. (4.19), i.e. $dU = TdS - PdV$, yields the following:

$$\left(\frac{\delta U}{\delta S}\right)_V = T \tag{4.27}$$

and

$$\left(\frac{\delta U}{\delta S}\right)_V = -P \tag{4.28}$$

Thus, by specifying U as a function of S and V, it is possible to evaluate the values of T and P, thereby describing the system fully.

Equation (4.27) indicates that in a reversible isochoric process, $\Delta U = \int T\mathrm{d}S = \int \delta q = q$, i.e. the change in internal energy is equal to the reversible heat exchanged.

Similarly, it is clear from Eq. (4.28) that in a reversible adiabatic or isentropic process, $\Delta U = -\int P\mathrm{d}V = -\int \delta w = -w$, i.e. the change in internal energy is equal to the reversible work done on the system.

4.3.2 Criteria of Equilibrium Under Constant Pressure and Entropy Conditions

Here, the conditions are $\mathrm{d}P = 0$ and $\mathrm{d}S = 0$. We know that $H = U + PV$.

On differentiation, $\mathrm{d}H = \mathrm{d}U + P\mathrm{d}V + V\mathrm{d}P = T\mathrm{d}S - P\mathrm{d}V + P\mathrm{d}V + V\mathrm{d}P = T\mathrm{d}S + V\mathrm{d}P$

$$\text{When } \mathrm{d}P = 0 \text{ and } \mathrm{d}S = 0, \text{ then } \mathrm{d}H = 0 \text{ or } (\delta H)_{S,P} = 0 \tag{4.29}$$

Equation (4.29) clarifies that the enthalpy becomes the necessary and sufficient criteria for the establishment of equilibrium under constant pressure and entropy conditions.

As discussed in Sect. 4.3.1, $\mathrm{d}U + P\mathrm{d}V < T\mathrm{d}S$ for an irreversible/natural process.

Under the specified conditions (i.e. $dP = 0$ and $dS = 0$), $dU + PdV < 0$ or $\mathrm{d}(U + PV) < 0$.

Or

$$\mathrm{d}H < 0 \text{ or } (\delta H_{\mathrm{SP}}) < 0 \text{ for an irreversible/natural process} \tag{4.30}$$

$$\text{Hence, } \mathrm{d}H > 0 \text{ for an unnatural process} \tag{4.31}$$

Therefore, the net result is $\mathrm{d}H <=> 0$ for natural, equilibrium and unnatural processes as the case may be. The above discussion clarifies that such natural processes tend to proceed in a direction in which the enthalpy decreases, and the state of equilibrium is described by Eq. (4.29). Thus, the enthalpy fulfils the basic characteristics of a potential, as outlined in Sect. 4.3. Hence, it qualifies for being called a thermodynamic potential under constant pressure and entropy conditions. Therefore, the enthalpy can be mathematically described as a function of P and S by the following expression:

$$H = f(S, P)$$

On differentiation, the enthalpy change can be written as:

$$dH = \left(\frac{\delta H}{\delta S}\right)_P dS + \left(\frac{\delta H}{\delta P}\right)_S dP \tag{4.32}$$

From Eq. (4.20), we also know that $dH = TdS + VdP$. A comparison of this expression with Eq (4.32) yields the following to important relationships:

$$\left(\frac{\delta H}{\delta S}\right)_P = T \tag{4.33}$$

and

$$\left(\frac{\delta H}{\delta P}\right)_S = V \tag{4.34}$$

This shows that by expressing *H* as a function of *P* and *S*, the remaining variables *T* and *V* can be readily found out and thus the system gets completely described.

4.3.3 Criteria of Equilibrium Under Constant Temperature and Volume Conditions

Mathematically, these conditions are written as $dT = 0$ and $dV = 0$.

From Eq. (4.2), $A = U - TS$.

Differentiation of this equation yields the following:

$$dA = dU - TdS - SdT = TdS - PdV - TdS - SdT = -PdV - SdT$$

For a reaction/process at constant temperature and volume, the above equation reduces to

$$dA = 0 \text{ or } (\delta A)_{V,T} = 0 \tag{4.35}$$

This indicates that Helmholtz free energy becomes the necessary and sufficient criteria for the attainment of equilibrium under isothermal and isochoric conditions.

The following relation has been outlined in Sect. 4.3.1: $dU + PdV < TdS$ for an irreversible/natural process or $dU - TdS < -PdV$ or $dU - TdS - SdT < -PdV - SdT$ or $d(U - TS) < -PdV - SdT$ or $dA < -PdV - SdT$ for an irreversible/natural process.

For natural processes, occurring under isothermal and isochoric conditions, this equation yields the following:

$$dA < 0 \tag{4.36}$$

On combining Eqs. (4.35) and (4.36), $\mathrm{d}A \leq 0$, which signifies that the change in Helmholtz free energy is either zero for a reversible/equilibrium process or less than zero for an irreversible/natural process. This leads to the conclusion that all natural processes, occurring at constant temperature and volume, are accompanied by a decrease in Helmholtz free energy of the system and the reversible process is described by $\mathrm{d}A = 0$. Hence, Helmholtz free energy qualifies for being referred as a thermodynamic potential for the processes conducted under constant temperature and volume conditions. Therefore, Helmholtz free energy is generally expressed as a function of temperature and volume as follows:

$$A = f(V, T) \tag{4.37}$$

On differentiation, Eq. (4.37) leads to the following expression:

$$\mathrm{d}A = \left(\frac{\delta A}{\delta V}\right)_T \mathrm{d}V + \left(\frac{\delta A}{\delta T}\right)_V \mathrm{d}T \tag{4.38}$$

From Eq. (4.21), we also know that $\mathrm{d}A = -P\mathrm{d}V - S\mathrm{d}T$. Now, comparison of this equation with Eq. (4.38) gives the following two relationships:

$$\left(\frac{\delta A}{\delta V}\right)_T = -P \tag{4.39}$$

$$\left(\frac{\delta A}{\delta T}\right)_V = -S \tag{4.40}$$

Thus, by specifying A as a function of V and T, the remaining variables P and S can be readily found out and the system gets completely described.

4.3.4 Criteria of Equilibrium Under Constant Temperature and Pressure Conditions

These are the most commonly adopted conditions in chemical and metallurgical processes and mathematically expressed as $\mathrm{d}T = 0$ and $\mathrm{d}P = 0$. As known to us, Gibbs free energy$G = H - TS$.

Differentiation of this equation yields the following:

$$\mathrm{d}G = \mathrm{d}H - T\mathrm{d}S - S\mathrm{d}T = T\mathrm{d}S + V\mathrm{d}P - T\mathrm{d}S - S\mathrm{d}T = V\mathrm{d}P - S\mathrm{d}T$$

On putting $\mathrm{d}T = 0$ and $\mathrm{d}P = 0$, it reduces to the following:

$$\mathrm{d}G = 0 \quad \text{or} \quad (\partial G)_{P,T} = 0 \tag{4.41}$$

This shows that Gibbs free energy becomes the necessary and sufficient criteria for the achievement of equilibrium under constant temperature and pressure conditions.

As outlined in Sect. 4.3.1, $\mathrm{d}U + P\mathrm{d}V < T\mathrm{d}S$ for an irreversible/natural process. Addition of the term $V\mathrm{d}P - T\mathrm{d}S - S\mathrm{d}T$ in both sides of this equation yields the following:

$$\mathrm{d}U + P\mathrm{d}V + V\mathrm{d}P - T\mathrm{d}S - S\mathrm{d}T < T\mathrm{d}S + V\mathrm{d}P - T\mathrm{d}S - S\mathrm{d}T$$

or $\mathrm{d}(U + PV - TS) < V\mathrm{d}P - S\mathrm{d}T$

or $\mathrm{d}(H - TS) < V\mathrm{d}P - S\mathrm{d}T$ or $\mathrm{d}G < V\mathrm{d}P - S\mathrm{d}T$ for an irreversible/natural process. Hence, for a reaction/process at constant temperature and pressure, this equation reduces to the following:

$$\mathrm{d}G < 0 \text{ for an irreversible/natural process} \tag{4.42}$$

Conversely and obviously, $\mathrm{d}G > 0$ for an unnatural processes.

On combining Eqs. (4.41) and (4.42), $\mathrm{d}G \leq 0$. This indicates that under the specified conditions (i.e. $\mathrm{d}T = 0$ and $\mathrm{d}P = 0$) the change in Gibbs free energy is either zero for a reversible/equilibrium process or less than zero for a natural process. This concludes that all natural/spontaneous processes, which occur at constant temperature and pressure, are accompanied by a decrease in Gibbs free energy of the system and the reversible process is described by $\mathrm{d}G = 0$. At equilibrium state, the free energy attains the minimum value. Hence, Gibbs free energy qualifies for being called a thermodynamic potential for the processes operated under constant temperature and pressure conditions. Therefore, Gibbs free energy is generally expressed as a function of temperature and pressure as follows:

$$G = f(P, T) \tag{4.43}$$

On differentiation, Eq. (4.43) leads to the following expression:

$$\mathrm{d}G = \left(\frac{\delta G}{\delta P}\right)_T \mathrm{d}P + \left(\frac{\delta G}{\delta T}\right)_P \mathrm{d}T \tag{4.44}$$

From Eq. (4.22), we also know that $\mathrm{d}G = -S\mathrm{d}T + V\mathrm{d}P$. A comparison of this equation with Eq. (4.44) leads to the following two relationships:

$$\left(\frac{\delta G}{\delta P}\right)_T = V \tag{4.45}$$

$$\left(\frac{\delta G}{\delta T}\right)_P = -S \tag{4.46}$$

Thus, by expressing G as a function of P and T, the remaining variables V and S can be readily calculated and the system gets completely described.

In this section, we have arrived at four thermodynamic potentials, viz. internal energy, enthalpy, Helmholtz free energy and Gibbs free energy. Internal energy and Helmholtz free energy have got more application in solid state processes where volume remains approximately constant. All these functions are extensive properties and their absolute values cannot be measured. Hence, during a process, changes in their values are generally determined.

4.4 Change in Free Energy and Feasibility of a Reaction

In order to study the change in concentration of reactants and products, knowledge of both the equilibrium condition and reaction rate is required. The free energy change of a reaction is a measure of the first of these. For application of ΔG criterion, the process has to be isothermal and isobaric. Whenever a set of reactants under a state of thermal and mechanical equilibrium (i.e. constant temperature and pressure) leads to the formation of a new set of products, there is a change in free energy of the system. This change in free energy is equal to the difference between the sum of free energies of the products (i.e. $\sum G_P$) and that of the reactants (i.e. $\sum G_R$) at constant temperature and pressure.

Let us consider the reaction $A + B \leftrightarrow AB$. The change free energy for a forward reaction is given by the following equation:

$$\begin{aligned}\Delta G &= \sum G_P - \sum G_R = G_{AB} - (G_A + G_B) \\ &= H_{AB} - TS_{AB} - (H_A - TS_A + H_B - TS_B) \\ &= (H_{AB} - H_A - H_B) - T(S_{AB} - S_A - S_B) \\ &= \Delta H - T\Delta S \end{aligned} \tag{4.47}$$

where ΔG is the available energy to do the work and $T\Delta S$ (i.e. $\Delta H - \Delta G$) represents the unavailable energy which is utilized to cause randomness in the system. In order to understand the feasibility of a reaction well, one will have to consider the three possibilities, such as (i)$\Delta G > 0$, (ii) $\Delta G < 0$ and (iii) $\Delta G = 0$.

Case I: $\Delta G > 0$ means $\sum G_P > \sum G_R$, i.e. the products have higher capacity to react or do the chemical work than the reactants. Hence, backward (i.e. reverse) reaction is feasible.

Case II: $\Delta G < 0$ means $\sum G_R > \sum G_P$, i.e. the reactants have higher capacity to react or do the chemical work than the products. Hence, forward reaction is feasible.

Case III: $\Delta G = 0$ means $\sum G_P = \sum G_R$, i.e. the reactants and products both have equal capacity to react or do the chemical work. This indicates no change in concentration of the either, i.e. the amounts of reactants and products will remain unchanged during the course of reaction and chemical equilibrium is said to be established in the system.

For a system involving simultaneous forward and backward reactions, the overall free energy change, ΔG_O, for the reaction is expressed by the following relationship:

$$\Delta G_O = \Delta G_f - \Delta G_b \quad (4.48)$$

where ΔG_f and ΔG_b represent the changes in free energy associated with forward and backward reactions.

Let us assume that the process consists only a change of pressure, i.e.

$$X\,(P_1, T) \rightarrow X\,(P_2, T)$$

If X is solid or liquid, $\Delta G = 0$, because change of pressure does not have much effect on its free energy. However, if X is a gas then, change of pressure would have a significant effect on G. From Eq. (4.22), at constant temperature, $\mathrm{d}G = V\mathrm{d}P$. $PV = \mathrm{RT}$ if X is an ideal gas.

Now,

$$\mathrm{d}G = V\mathrm{d}P = \left(\frac{\mathrm{RT}}{P}\right)\,\mathrm{d}P = \mathrm{RTd}\ln P \quad (4.49)$$

On integrating this equation within the limits P_1 to P_2, we have the following:

$$\Delta G = \mathrm{RT}\ln\left(\frac{P_2}{P_1}\right) \quad (4.50)$$

The application of Eq. (4.49) is restricted only to an ideal gas. In order to apply this equation to any thermodynamic substances (whether ideal or non-ideal), the term P needs to be replaced by a function f (i.e. fugacity) and this equation gets modified to $\mathrm{d}G = \mathrm{RTd}\ln f$ for an infinitesimal change or $\Delta G = \mathrm{RT}\ln f$ for a finite change.

4.4.1 ΔH and ΔS Values, and Feasibility of a Reaction

In this section, feasibility of chemical reaction in context to different values of ΔH and $\Delta \mathrm{S}$, as presented below, has been examined by using the relation $\Delta G = \Delta H - T\Delta S$:

Case I: If $\Delta H = 0$ (i.e. adiabatic condition) and $\Delta S > 0$ (i.e. irreversible/natural/spontaneous process), then $\Delta G = -T\Delta S$, i.e. $\Delta G < 0$ or negative and thus, forward reaction is feasible.

Case II: If $\Delta H > 0$ (i.e. endothermic reaction) and $\Delta S = 0$ (i.e. reversible condition), then $\Delta G = \Delta H$, i.e. $\Delta G > 0$ or positive and thus, a reaction is feasible in backward/reverse direction.

Case III: If $\Delta H < 0$ (i.e. exothermic reaction) and $\Delta S = 0$, then $\Delta G = -\Delta H$, i.e. $\Delta G < 0$ and this favours the reaction in forward direction.

Case IV: If $\Delta H < 0$ and $\Delta S > 0$, then $\Delta G = -\Delta H - T\Delta S$, i.e. $\Delta G < 0$. Hence, a reaction in forward direction is feasible.

Case V: If $\Delta H < 0$ and $\Delta S < 0$, then $\Delta G = -\Delta H + T\Delta S$. At low temperatures, $T\Delta S < \Delta H$ and $\Delta G < 0$. Hence, a forward reaction is feasible in this case. At high temperatures, $T\Delta S > \Delta H$ and $\Delta G > 0$. This favours the reaction in backward/reverse direction.

Case VI: If $\Delta H > 0$ and $\Delta S > 0$, then $\Delta G = \Delta H - T\Delta S$. At low temperatures, $T\Delta S < \Delta H$ and $\Delta G > 0$. This means that the low temperatures under the above specified conditions favour the reaction in reverse direction. On the other hand, at high temperatures, $T\Delta S > \Delta H$ and $\Delta G < 0$. This indicates that the high temperatures under the above-mentioned conditions promote the reaction in forward direction.

Case VII: If $\Delta H > 0$ and $\Delta S < 0$, then $\Delta G = \Delta H + T\Delta S$, i.e. $\Delta G > 0$. Hence, the above specified conditions force the reaction to proceed in reverse direction.

4.5 Standard States

Standard state of materials means the state in which they are stable at the specified temperature and one atmospheric pressure. For solids and liquids, the standard state, if not stated, is normally their pure and most stable forms at the specified temperature and one atmospheric pressure. The activity of a material in its pure form is assumed to be unity and that of the impure form is expressed in fraction. Alternatively, the standard state of a material is defined as the form in which its activity is taken as unity (i.e. $a = 1$) at the specified temperature and one atmospheric pressure. By definition, activity is a numerical fraction to express the interactive ability of the material with respect to some standard interactive ability. The interactive ability is maximum when the material is in its pure form, i.e. $a = 1$. This pure form can be solid, liquid or gas of that substance under the given conditions of temperature and pressure. Gaseous constituents are considered to be in their standard states if their partial pressures are one atmosphere and unit fugacity (i.e. $f = 1$) at the specified temperature. This has also been discussed in Sect. 5.1.1.

At one atmospheric pressure, the stable state of H_2O is an ice at a temperature below 0 °C, liquid water from 0 to 100 °C, and a gas at temperatures above 100 °C. Another example of interest is the allotropic transformations in pure iron. The standard states of an iron at temperatures of 973, 1400 and 1700 K and one atmospheric pressure are α-Fe, γ-Fe and δ-Fe. By convention, the thermodynamic properties (e.g. *H*, *G*, *U*, etc.) at standard state are denoted by superscript "0", e.g. H^0, G^0, etc. the choice of standard state is purely arbitrary and without any scientific basis, though perfectly useful.

4.6 Dependence of ΔG on Temperature and Pressure

ΔG criterion is useful in predicting equilibrium and spontaneity of the process only at constant temperature and pressure. Majority of the chemical and metallurgical processes/reactions are carried out under constant pressure condition. However, one has to analyse many problems (viz. geological, etc.) taking place over a wide range of temperatures and pressures. Hence, derivation of equations showing the relationships between ΔG, T and P is essential.

From Eq. (4.22), we know that $\mathrm{d}G = V\mathrm{d}P - S\mathrm{d}T$. Thus, the total differential for ΔG can be written as follows:

$$\mathrm{d}\Delta G = \Delta V\mathrm{d}P - \Delta S\mathrm{d}T \tag{4.51}$$

Let the temperature of the system varies from 298 (i.e. reference temperature) to T K and pressure from P^0 to P. Integration of Eq. (4.51) within these temperatures and pressure limits yields the following:

$$\int_{298,P^0}^{T,P} \mathrm{d}\Delta G = \int_{T,P^0}^{T,P} \Delta V\mathrm{d}P + \int_{298,P^0}^{T,P^0} (-\Delta S)\mathrm{d}T \tag{4.52}$$

It is clear from this equation that the expression having $\mathrm{d}P$ term in right side is integrated with respect to pressure at constant temperature T and the expression consisting of $\mathrm{d}T$ term in the same side is integrated with respect to temperature at constant pressure P^0. Hence, the dependence of ΔG on temperature and pressure can be dealt with independently.

4.6.1 Derivation of an Equation for ΔG Versus T Relation

From the ΔH and ΔS values usually easily available at reference temperature (i.e. 298 K), the free energy change associated with a process or reaction can be calculated directly at that temperature by using the formula $\Delta G_{298} = \Delta H_{298} - 298\Delta S_{298}$. For calculation of ΔG at other temperature, the variation of ΔH and ΔS must be considered and the free energy change at any temperature T can be obtained by using the following equation:

$$\Delta G_T = \Delta H_T - T\Delta S_T \tag{4.53}$$

Based on the expressions, as outlined in Sects. (2.4.2) and (3.6.5), for the variation enthalpy and entropy with temperature, ΔH and ΔS of the reaction/process at any temperature T can be expressed by the following equations for the temperature range of 0 to T K:

$$\Delta H_T = \Delta H_0 + \int_0^T \Delta C_P \mathrm{d}T \tag{4.54}$$

And

$$\Delta S_T = \Delta S_0 + \int_0^T \frac{\Delta C_P}{T} \mathrm{d}T \tag{4.55}$$

where ΔH_0 and ΔS_0 are the enthalpy and entropy changes at 0 K, and ΔC_P is the difference in the specific heats of products and reactants. With the help of these expressions of ΔH_T and ΔS_T, Eq. (4.53) can be written in the following form:

$$\Delta G_T = \Delta H_0 + \int_0^T \Delta C_P \mathrm{d}T - T\left[\Delta S_0 + \int_0^T \frac{\Delta C_P}{T} \mathrm{d}T\right] \tag{4.56}$$

If the considered temperature range in the above derivation is from 298 to T K, and the reactants and the products are in their standard states (i.e. pure and at one atmospheric pressure), then Eq. (4.56) gets modified as follows:

$$\Delta G_T^0 = \Delta H_{298}^0 + \int_{298}^T \Delta C_P^0 \mathrm{d}T - T\left[\Delta S_{298}^0 + \int_{298}^T \frac{\Delta C_P^0}{T} \mathrm{d}T\right] \tag{4.57}$$

Equations (4.56) and (4.57) represent the variation of free energy change with temperature in the range 0 to T K and 298 to T K. Equation (4.57) indicates that ΔG_T^0 at any higher temperature can be evaluated by using the ΔH^0 and ΔS^0 data at 298 K and Kirchhoff's law. If $\Delta C_P^0 = 0$, then both ΔH_T^0 and ΔS_T^0 become independent of temperature and we get,

$$\Delta G_T^0 = \Delta H_{298}^0 - T\Delta S_{298}^0$$

When one or more phase transformations occur in the considered temperature range, the expressions for ΔH_T and ΔS must be expanded in Eq. (4.56) to include the appropriate enthalpies and entropies of transformations, as indicated in Chaps. 2 and 3.

4.6.2 Derivation of an Equation for ΔG Versus P Relation

On the basis of expressions (4.45) and (4.46), as outlined in Sect. 4.3.4, we may write the followings:

$$\left(\frac{\delta \Delta G}{\delta P}\right)_T = \Delta V \tag{4.58}$$

and

$$\left(\frac{\delta \Delta G}{\delta T}\right)_P = -\Delta S \tag{4.59}$$

From Eq. (4.58),

$$\mathrm{d}(\Delta G) = \Delta V \mathrm{d}P \text{ at constant temperature } T \tag{4.60}$$

Let the pressure of the system changes from P^0 to P at a constant temperature T. on integration of Eq. (4.60) within this pressure limit, we get the following:

$$\int_{P^0}^{P} \mathrm{d}\Delta G = \int_{P^0}^{P} \Delta V \mathrm{d}P \quad \text{or} \quad \Delta G(\text{at } P) = \Delta G\left(\text{at } P^0\right) + \int_{P^0}^{P} \Delta V \mathrm{d}P \tag{4.61}$$

ΔV must be known for the temperature at which an integration is carried out.

General Expression: on combining Eqs. (4.57) and (4.61), we obtain the general expression for the dependence of ΔG on T and P as follows:

$$\Delta G^0(P, T) = \Delta H^0_{298} - T \Delta S^0_{298} + \int_{298}^{T} \Delta C^0_P \mathrm{d}T - T \int_{298}^{T} \frac{\Delta C^0_P}{T} \mathrm{d}T + \int_{P^0}^{P} \Delta V \mathrm{d}P \tag{4.62}$$

It is clear from this equation that in order to calculate free energy change at any temperature and pressure, we need to know the values of ΔH and ΔS at standard conditions (i.e. at 298 K and atmospheric pressure), ΔC_P as a function of temperature, and the value of ΔV as a function of pressure at temperature T. If the data of ΔC_P and ΔV are not available, an approximate result can be obtained as follows by assuming them constant over the considered range:

$$\begin{aligned} \Delta G^0_{\text{approx}}(P, T) = {} & \Delta H^0_{298} - T \Delta S^0_{298} + \Delta C^0_P (T - 298) \\ & - T \Delta C^0_P (\ln T - \ln 298) + \Delta V\left(P - P^0\right) \end{aligned} \tag{4.63}$$

4.7 Gibbs–Helmholtz Equations

Gibbs–Helmholtz equations give relation between ΔG, ΔH and T. These equations are very frequently use for the determination of free energy change as a function of temperature during a process from the data obtained by calorimetric measurements. For derivation of these equations, mass, composition and pressure are assumed to be constant and only the temperature change is considered. Let us start with Eq. (4.22) derived for the combined statement of 1st and 2nd laws of thermodynamics which as follows:

$$\mathrm{d}G = V\mathrm{d}P - S\mathrm{d}T$$

At constant pressure, $\mathrm{d}P = 0$ and we have the following:

$$-S = \frac{\mathrm{d}G}{\mathrm{d}T} = \left(\frac{\partial G}{\partial T}\right)_P \tag{4.64}$$

Substitution of the value of S from Eq. (4.64) in the expression, $G = H - TS$ leads to the following:

$$G = H - TS = H + T\left(\frac{\partial G}{\partial T}\right)_P \tag{4.65}$$

On dividing both sides by T^2, we get the following expression:

$$\frac{G}{T^2} = \frac{H}{T^2} + \frac{1}{T}\left(\frac{\partial G}{\partial T}\right)_P$$

or

$$\frac{1}{T}\left(\frac{\partial G}{\partial T}\right)_P - \frac{G}{T^2} = -\frac{H}{T^2}$$

or

$$\left[\frac{\partial(G/T)}{\partial T}\right]_P = -\frac{H}{T^2} \tag{4.66}$$

Further rearrangement of the terms leads to the following:

or

$$\left[\frac{\partial(G/T)}{\partial(1/T)}\right]_P = H \tag{4.67}$$

Equations (4.66) and (4.67) are the two different forms of Gibbs–Helmholtz equation for the variation of Gibbs free energy with temperature during a chemical or physical change.

Now, let us considered the following general chemical reaction:

$$xX + yY = mM + nN$$

where x, y, m and n are the numbers of moles of constituents X, Y, M and N taking part in the chemical reaction. The free energy change (i.e. ΔG) for the above chemical reaction can be written as follows:

$$\Delta G = \sum G_{\text{products}} - \sum G_{\text{reactants}} = (mG_M + nG_N) - (xG_X + yG_Y) \quad \text{or}$$

$$\frac{\Delta G}{T} = \left(m\frac{G_M}{T} + n\frac{G_N}{T}\right) - \left(x\frac{G_X}{T} + y\frac{G_Y}{T}\right)$$

On differentiation with respect to $1/T$ at constant pressure, the above expression becomes as follows:

$$\left[\frac{\partial(\Delta G/T)}{\partial(1/T)}\right]_P = m\left[\frac{\partial(G_M/T)}{\partial(1/T)}\right]_P + n\left[\frac{\partial(G_N/T)}{\partial(1/T)}\right]_P - x\left[\frac{\partial(G_X/T)}{\partial(1/T)}\right]_P - y\left[\frac{\partial(G_Y/T)}{\partial(1/T)}\right]_P$$

$$\left[\frac{\partial(\Delta G/T)}{\partial(1/T)}\right]_P = mH_M + nH_N - xH_X - yH_Y = \Delta H$$

Thus, for a chemical reaction or physical change, one can write the Gibbs–Helmholtz equation in the following form:

$$\left[\frac{\partial(\Delta G/T)}{\partial(1/T)}\right]_P = \Delta H \tag{4.68}$$

When the reactants and the products are in their standard states, then the above equation can be written as follows:

$$\left[\frac{\partial(\Delta G^o/T)}{\partial(1/T)}\right]_P = \Delta H^o \tag{4.69}$$

where ΔG represents the change in Gibbs free energy for a process and ΔH refers to change in enthalpy.

For derivation of an alternate form of Gibbs–Helmholtz equation in terms of variation of Helmholtz free energy at constant volume with temperature, one can start with Eq. (4.21), i.e. $\mathrm{d}A = -S\mathrm{d}T - P\mathrm{d}V$.

At constant volume, $\mathrm{d}V = 0$ and $\left(\frac{\delta A}{\delta T}\right)_V = -S$. Substitution of this value of S in the expression $A = U - TS$ yields the following:

$$A = U - TS = U + T\left(\frac{\delta A}{\delta T}\right)_V$$

Now, the same derivation procedure is followed, as done in developing Eqs. (4.66) to (4.69), which leads to the development of another form of Gibbs–Helmholtz equation in terms of Helmholtz free energy and internal energy as follows:

$$\left[\frac{\partial(A/T)}{\partial(1/T)}\right]_V = U \quad (4.70)$$

For a chemical reaction or physical change, the above expression can be written as follows:

$$\left[\frac{\partial(\Delta A/T)}{\partial(1/T)}\right]_V = \Delta U \quad (4.71)$$

For standard state condition, this equation can be written in the following form:

$$\left[\frac{\partial(\Delta A^o/T)}{\partial(1/T)}\right]_V = \Delta U^o \quad (4.72)$$

where ΔU is the change in internal energy during a process. Majority of the chemical and metallurgical processes are carried out under constant pressure condition. Hence, Eqs. (4.70) to (4.72) are less frequently used than Eqs. (4.66) to (4.69).

4.8 Maxwell's Relations

In a single-component system of fixed composition, the specification of two state variables completely determines the state of the system. Maxwell's relation are a set of equations in thermodynamics which are derived from the definitions of the thermodynamic potentials (i.e. U, H, A, G, etc.). If Z is a state function and x and y are the chosen independent state variables (may be T and P or others) for a system of fixed composition, then

$$Z = f(x, y)$$

On differentiation gives the following:

$$\mathrm{d}Z = \left(\frac{\partial Z}{\partial x}\right)_y \mathrm{d}x + \left(\frac{\partial Z}{\partial y}\right)_x \mathrm{d}y = M\mathrm{d}x + N\mathrm{d}y$$

where $M = \left(\frac{\partial Z}{\partial x}\right)_y$ and $N = \left(\frac{\partial Z}{\partial y}\right)_x$. M and N both are, in general, functions of x and y.

Now,

$$\left(\frac{\partial Z}{\partial x}\right)_y = M$$

On differentiation with respect to y yields the following:

$$\left[\frac{\partial}{\partial y}\left(\frac{\partial Z}{\partial x}\right)_y\right]_x = \left(\frac{\partial M}{\partial y}\right)_x$$

Similarly, $\left[\frac{\partial}{\partial x}\left(\frac{\partial Z}{\partial y}\right)_x\right]_y = \left(\frac{\partial N}{\partial x}\right)_y$.

As Z is a state function, the change in Z is independent of the order of differentiation, i.e.

$$\left[\frac{\partial}{\partial y}\left(\frac{\partial Z}{\partial x}\right)_y\right]_x = \left[\frac{\partial}{\partial x}\left(\frac{\partial Z}{\partial y}\right)_x\right]_y = \frac{\partial^2 Z}{\partial x \partial y}$$

Hence,

$$\left(\frac{\partial M}{\partial y}\right)_x = \left(\frac{\partial N}{\partial x}\right)_y \tag{4.73}$$

In Eqs. (4.19) to (4.22), we have already expressed the thermodynamic potentials (i.e. U, H, A and G) in the following differential forms:

$$\mathrm{d}U = T\mathrm{d}S - P\mathrm{d}V \quad \mathrm{d}H = T\mathrm{d}S + V\mathrm{d}P$$
$$\mathrm{d}A = -S\mathrm{d}T - P\mathrm{d}V \text{ and } \mathrm{d}G = -S\mathrm{d}T + V\mathrm{d}P$$

Thus, one can say that the natural variables of U, H, A and G are S and V (both extensive), S and P (mixed, i.e. extensive and intensive), V and T (mixed), and P and T (both intensive). Equation (4.73) can be applied to Eqs. (4.19) to (4.22) because $\mathrm{d}U$, $\mathrm{d}H$, $\mathrm{d}A$ and $\mathrm{d}G$ are exact differentials and this application leads to the development of following relationships:

From Eq. (4.19),

$$\left(\frac{\partial T}{\partial V}\right)_S = -\left(\frac{\partial P}{\partial S}\right)_V \tag{4.74}$$

From Eq. (4.20),

$$\left(\frac{\partial T}{\partial P}\right)_S = \left(\frac{\partial V}{\partial S}\right)_P \tag{4.75}$$

From Eq. (4.21),

$$\left(\frac{\partial S}{\partial V}\right)_T = \left(\frac{\partial P}{\partial T}\right)_V \tag{4.76}$$

From Eq. (4.22),

$$\left(\frac{\partial S}{\partial P}\right)_T = -\left(\frac{\partial V}{\partial T}\right)_P \tag{4.77}$$

Equations (4.74) to (4.77) are popularly known as Maxwell's relations and have been found to be very valuable in science and engineering. The latter two equations, i.e. (4.76) and (4.77), are particularly more useful in dealing with the thermodynamic aspects of the processes. With the help of these equations, the variation of S with (i) V and (ii) P at constant temperature can be determined from knowledge of the variation of P with T at constant volume and V with T at constant pressure. By combining these four basic equations with others, it is possible to derive many other thermodynamic relations in the form of differential equations.

4.8.1 Application of Maxwell's Relations: Few Examples

Some examples showing the use of Maxwell's equations are discussed below as follows:

(a) For a closed system of fixed composition, Eq. (4.19) gives $\mathrm{d}U = T\mathrm{d}S - P\mathrm{d}V$.

On dividing both sides of this equation by $\mathrm{d}V$ and considering the process to be an isothermal, the above equation becomes as:

$$\left(\frac{\partial U}{\partial V}\right)_T = T\left(\frac{\partial S}{\partial V}\right)_T - P \tag{4.78}$$

From Maxwell's Eq. (4.76), we know that

$$\left(\frac{\partial S}{\partial V}\right)_T = \left(\frac{\partial P}{\partial T}\right)_V$$

Use of this Maxwell's relation in Eq. (4.78) yields the following:

$$\left(\frac{\partial U}{\partial V}\right)_T = T\left(\frac{\partial P}{\partial T}\right)_V - P \tag{4.79}$$

This equation relates the internal energy U of a closed one component system to the experimentally measurable quantities T, P and V.

If the system is an ideal gas, i.e., $PV = \text{RT}$, then $(\delta P/\delta T)_V = R/V$ and Eq. (4.79) gets modified to the following:

$$\left(\frac{\delta U}{\delta V}\right)_T = \frac{RT}{V} - P = P - P = 0$$

This indicates that the internal energy of an ideal gas at constant temperature is independent of the volume of the gas.

(b) For a closed system of fixed composition, Eq. (4.20) gives $\mathrm{d}H = T\mathrm{d}S + V\mathrm{d}P$.

On dividing both sides of this equation by $\mathrm{d}p$ and considering temperature to be constant, the above equation becomes as:

$$\left(\frac{\partial H}{\partial P}\right)_T = T\left(\frac{\partial S}{\partial P}\right)_T + V \tag{4.80}$$

From Maxwell's Eq. (4.77), we know that

$$\left(\frac{\partial S}{\partial P}\right)_T = -\left(\frac{\partial V}{\partial T}\right)_P$$

Use of this Maxwell's relation allows Eq. (4.80) to be written as follows:

$$\left(\frac{\partial H}{\partial P}\right)_T = -T\left(\frac{\partial V}{\partial T}\right)_P + V \tag{4.81}$$

This equation relates the enthalpy H of a closed system to the experimentally measurable quantities T, P and V.

If again the system is an ideal gas, then $(\partial V/\partial T)_P = R/P$ and Eq. (4.81) gets modified to the following:

$$\left(\frac{\partial H}{\partial P}\right)_T = -\frac{RT}{P} + V = -V + V = 0$$

This indicates that the enthalpy of an ideal gas at constant temperature is independent of the pressure of the gas.

(c) As outlined in Sect. 2.3.1, we have the following relationship between C_P and C_V:

$$C_P - C_V = \left(\frac{\partial V}{\partial T}\right)_P \left[P + \left(\frac{\partial U}{\partial V}\right)_T\right]$$

Equation (4.21) gives $dA = -SdT - PdV$, at constant temperature, $dT = 0$ and $dA = -PdV$.

Hence, $P = -\left(\frac{\partial A}{\partial V}\right)_T$. Thus, the above equation becomes as follows:

$$C_P - C_V = \left(\frac{\partial V}{\partial T}\right)_P \left[\left(\frac{\partial U}{\partial V}\right)_T - \left(\frac{\partial A}{\partial V}\right)_T\right] \tag{4.82}$$

From Eq. (4.2), we know that $A = U - TS$.

On differentiation of this equation with respect to V at constant temperature, we get the following:

$$\left(\frac{\partial A}{\partial V}\right)_T = \left(\frac{\partial U}{\partial V}\right)_T - T\left(\frac{\partial S}{\partial V}\right)_T \tag{4.83}$$

Combination of Eqs. (4.82) and (4.83) yields the following:

$$C_P - C_V = \left(\frac{\partial V}{\partial T}\right)_P \left[\left(\frac{\partial U}{\partial V}\right)_T - \left(\frac{\partial U}{\partial V}\right)_T + T\left(\frac{\partial S}{\partial V}\right)_T\right] = T\left(\frac{\partial V}{\partial T}\right)_P \left(\frac{\partial S}{\partial V}\right)_T \tag{4.84}$$

After the use of Maxwell's relation outlined in Eqs. (4.76), (4.84) becomes as follows:

$$C_P - C_V = T\left(\frac{\partial V}{\partial T}\right)_P \left(\frac{\partial P}{\partial T}\right)_V \tag{4.85}$$

By the use of Eq. (1.4) in $V = f(P, T)$, we obtain the following:

$$\left(\frac{\partial V}{\partial P}\right)_T \left(\frac{\partial P}{\partial T}\right)_V \left(\frac{\partial T}{\partial V}\right)_P = -1 \tag{4.86}$$

Combination of Eqs. (4.85) and (4.86) leads to the following:

$$C_P - C_V = -T\left(\frac{\partial V}{\partial T}\right)_P \left(\frac{\partial P}{\partial V}\right)_T \left(\frac{\partial V}{\partial T}\right)_P \tag{4.87}$$

Now, isobaric coefficient of volumetric thermal expansion or isobaric thermal expansivity of the material (α) is defined by the following equation:

$$\alpha = \frac{1}{V}\left(\frac{\partial V}{\partial T}\right)_P \tag{4.88}$$

Similarly, isothermal compressibility of a material (β) is defined by the equation as follows:

$$\beta = -\frac{1}{V}\left(\frac{\partial V}{\partial P}\right)_T \tag{4.89}$$

The symbol β denotes the decrease in volume of the system for unit increase of pressure acting on the system at constant temperature. Negative sign is used to make β a positive number.

Combination of Eqs. (4.87) to (4.89) gives the following expression:

$$C_P - C_V = \frac{VT\alpha^2}{\beta} \tag{4.90}$$

In case of solids, use of equation of state poses numerous problems. Hence, instead of using the equation of state, one determines experimentally α and β properties of materials. These two properties along with the data on C_P form the set of experimentally determinable variables from which changes in thermodynamic potential can be calculated. The compressibility of a stable phase is always positive (i.e. $\beta > 1$). This condition is generally referred to as the condition for mechanical stability.

4.9 Clausius–Clapeyron Equation

The equilibrium vapour pressures of metals are of great interest in metallurgy. When metals are kept at high temperatures for a longer period of time, there is a substantial loss of metals in the form of vapours. For example, in LD process of steel making, about 1% of iron gets lost in the vapour form and causes pollution in atmosphere. To minimize such losses, conditions need to be generated to reduce vaporization of metals and this requires knowledge of equilibrium vapour pressures of metals over a range of temperature and pressure of interest. Clausius–Clapeyron equation was developed to address this type of study. This equation enables us to quantify the effects of change of temperature and pressure on the melting points of solids, boiling points of liquids and any phase transformation.

Consider phase transformation of an element/compound from phase I to phase II. For equilibrium to be maintained at the phase transformation temperature and pressure,

$$\Delta G(\text{ph.I} \rightarrow \text{Ph.II}) = 0 \tag{4.91}$$

For an infinitesimal change in T and P, the free energies of both the phases change by a very small amount of $\mathrm{d}G(ph.I)$ and $\mathrm{d}G\ (ph.II)$. For the new equilibrium to be established between these two phases,

$$\{G(\text{ph.II}) + \mathrm{d}G(\text{ph.II})\} - \{G(\text{ph.I}) + \mathrm{d}G(\text{ph.I})\} = 0 \tag{4.92}$$

On combining Eqs. 4.91 and 4.92,

$$\mathrm{d}G(\text{ph.I}) = \mathrm{d}G(\text{ph.II}) \tag{4.93}$$

As outlined in Eq. 4.22, $\mathrm{d}G = -S\mathrm{d}T + V\mathrm{d}P$

Accordingly, $\mathrm{d}G(\text{ph.I}) = -S(\text{ph.I})\mathrm{d}T + V(\text{ph.I})\mathrm{d}P$ and $\mathrm{d}G(\text{ph.II}) = -S(\text{ph.II})\mathrm{d}T + V(\text{ph.II})\mathrm{d}P$.

From Eq. 4.93, we get $-S(\text{ph.I})\mathrm{d}T + V(\text{ph.I})\mathrm{d}P = -S(\text{ph.II})\mathrm{d}T + V(\text{ph.II})\mathrm{d}P$ or $\mathrm{d}P\{V(\text{ph.II}) - V(\text{ph.I})\} = \mathrm{d}T\{S(\text{ph.II}) - S(\text{ph.I})\}$

$$\left(\frac{\mathrm{d}P}{\mathrm{d}T}\right)_{\text{eq}} = \frac{S(\text{ph.II}) - S(\text{ph.I})}{V(\text{ph.II}) - V(\text{ph.I})} = \frac{\Delta S_{\text{tr}}}{\Delta V_{\text{tr}}} \tag{4.94}$$

where ΔS_t and ΔV_{tr} are the entropy and volume changes during transformation from phase I to phase II. At equilibrium during phase transformation, $\Delta G_n = 0$ or $\Delta G_{\text{tr}} = \Delta H_{\text{tr}} - T\Delta S_{\text{tr}} = 0$ or $\Delta S_{\text{tr}} = \Delta H_{\text{tr}}/T_{\text{txx}}$.

Therefore, Eq. 4.94 can be written as follows:

$$\left(\frac{\mathrm{d}P}{\mathrm{d}T}\right)_{\text{eq}} = \frac{\Delta H_{\text{tr}}}{T_{\text{tr}}\Delta V_{\text{tr}}} \tag{4.95}$$

where ΔH_{tr} is the enthalpy change during transformation from phase I to phase II. This equation is known as the Clausius–Clapeyron equation.

4.9.1 Application of Clausius–Clapeyron Equation

Clausius–Clapeyron equation is applicable to any phase change, such as.

(i) Melting/fusion or solid–liquid phase transformation equilibrium.
(ii) Allotropic transformation or solid–solid phase transformation equilibrium.
(iii) Vaporization or liquid–vapour phase transformation equilibrium.
(iv) Sublimation or solid–vapour phase transformation equilibrium.

Melting/fusion or solid–liquid phase transformation equilibrium: Clausius–Clapeyron equation for melting/fusion of a pure element or compound may be written as:

$$\left(\frac{\mathrm{d}P}{\mathrm{d}T}\right)_{\text{eq}} = \frac{\Delta H_m}{T_m \Delta V_m} = \frac{\Delta H_m}{T_m(V_L - V_S)}$$

Small change of temperature has negligible effect on ΔH_m and ΔV_m. Hence, they are taken as constant. Now,

$$\mathrm{d}P = \left[\frac{\Delta H_m}{(V_L - V_S)}\right]\frac{\mathrm{d}T}{T_m}$$

Let the melting point of solid changes from T_1 to T_2 upon increasing the pressure from P_1 to P_2. On integration between these limits, we have the following:

$$\int_{P_1}^{P_2} \mathrm{d}P = \frac{\Delta H_m}{(V_L - V_S)} \int_{T_1}^{T_2} \frac{\mathrm{d}T}{T}$$

$$\text{or} \quad P_2 - P_1 = \frac{\Delta H_m}{(V_L - V_S)} \cdot \ln \frac{T_2}{T_1} \tag{4.96}$$

Thus, the change in melting point of a solid with change of pressure may be calculated by use of this equation.

Allotropic transformation or solid–solid phase transformation equilibrium:

Let T is the transition temperature, α and β and are the stable phases above and below this temperature. On applying the Clausius–Clapeyron equation (4.95) to this solid–solid transformation, we get the following:

$$\left(\frac{\mathrm{d}P}{\mathrm{d}T}\right)_{\mathrm{eq}} = \frac{\Delta H_{\mathrm{tr}}}{T \Delta V_{\mathrm{tr}}}$$

where ΔH_{tr} is the amount of heat energy exchanged during this phase transformation.

$$\text{or} \quad \left(\frac{\mathrm{d}P}{\mathrm{d}T}\right)_{\mathrm{eq}} = \frac{\Delta H_{\mathrm{tr}}}{T(V_\alpha - V_\beta)}$$

where V_α and V_β represent the volumes of α and β phases. Upon integration,

$$P = \frac{\Delta H_{\mathrm{tr}}}{(V_\alpha - V_\beta)} \cdot \ln T + C$$

where C is the integration constant.

Vaporization or liquid–vapour phase transformation equilibrium: Transformation of liquid into its vapour is known as vaporization. The volume of vapour (V_{vp}) of a substance is much larger than that of liquid/solid phase $(V_{\mathrm{liq}}/V_{\mathrm{sol}})$. Hence, the volume change during vaporization,

$$\Delta V_{\mathrm{vp}} = V_{\mathrm{vp}} - V_{\mathrm{liq}} \approx V_{\mathrm{vp}}$$

From Eq. 4.95, the equation for vaporization can be written as

$$\left(\frac{\mathrm{d}P}{\mathrm{d}T}\right)_{\mathrm{eq}} = \frac{\Delta H_{\mathrm{vap}}}{\mathrm{TV}_{\mathrm{vap}}} \tag{4.97}$$

where ΔH_{vap} is the heat of vaporization. Usually, the vapour pressures are very low and such vapours can be assumed to behave as an ideal gas. Hence,

$$V_{vp} = \mathrm{RT}/P \tag{4.98}$$

where P is the vapour pressure of the substance. Combination of Eqs. (4.97) and (4.98) yields:

$$\left(\frac{\mathrm{d}P}{\mathrm{d}T}\right)_{\mathrm{eq}} = \frac{P\Delta H_{\mathrm{vap}}}{\mathrm{RT}^2}$$

$$\text{or}\quad \frac{1}{P}\left(\frac{\mathrm{d}P}{\mathrm{d}T}\right)_{\mathrm{eq}} = \frac{\Delta H_{\mathrm{vap}}}{\mathrm{RT}^2}$$

$$\text{or}\quad \mathrm{d}\ln P = \frac{\Delta H_{\mathrm{vap}}}{R}\cdot\frac{\mathrm{d}T}{T^2} \tag{4.99}$$

This is another form of Clausius–Clapeyron equation for liquid–vapour equilibrium. This equation correlates the vapour pressure with temperature. The values of all the ΔH_{vap}, R and T quantities are positive. Hence, the vapour pressure of a substance increases with rise of temperature and vice versa. Here, there are two possibilities:

Case I: Integration of Eq. (4.99) without taking limits of P and T.

Let us assume ΔH_{vap} to be constant. In reality, this is not true and the value of ΔH_{vap} varies with temperature. Integration of Eq. (4.99) yields the following:

$$\int \mathrm{d}\ln P = \frac{\Delta H_{\mathrm{vcp}}}{R}\int\frac{\mathrm{d}T}{T^2}\quad\text{or}$$

$$\ln P = -\frac{\Delta H_{\mathrm{vcp}}}{\mathrm{RT}} + C = -\frac{L_{\mathrm{vcp}}}{\mathrm{RT}} + C$$

where C is an integration constant and L_{vcp} is the latent heat of vaporization. This equation is of the form of a straight line equation. Hence, the plot between ln$P$ and $1/T$ will be a straight line, and slope of this line $= -\frac{\Delta H_{\mathrm{vap}}}{R}$ or $\Delta H_{\mathrm{vap}} = -(\text{slope}) \times R$.

Thus, the value of ΔH_{vap} can be calculated. Intercept of this line with ln$P$ axis will yield the value of integration constant C.

Case II: Integration of Eq. (4.99) by taking limits of P and T.

Let temperature of the system changes from T_1 to T_2 and the corresponding change in pressure is from P_1 to P_2. On assuming the considered temperature range to be small, the value of ΔH_{vap} will be independent of temperature. Then, on integration Eq. (4.99) between these limits, one can get the following:

$$\int_{R}^{P_2} \ln P = \frac{\Delta H_{\mathrm{vap}}}{R}\int_{T_1}^{T_2}\frac{\mathrm{d}T}{T^2}\quad\text{or}\quad \ln P_2 - \ln P_1 = -\frac{\Delta H_{\mathrm{vap}}}{R}\left(\frac{1}{T_2} - \frac{1}{T_1}\right) \tag{4.100}$$

By using this equation, one can calculate the vapour pressure at any temperature if the values of vapour pressure at other temperature and ΔH_{vap} over the considered temperature range are known.

Sublimation or solid–vapour phase transformation equilibrium: The transformation of a solid directly into its vapour is called sublimation. Basic equations for both the vaporization from liquid and solid (sublimation) are similar and the Clausius–Clapeyron equation for sublimation may be written as follows:

$$\mathrm{d}\ln P = \frac{\Delta H_{\text{sub}}}{R} \cdot \frac{\mathrm{d}T}{T^2}$$

On integration of this equation yields

$$P = -\frac{\Delta H_{\text{sub}}}{\text{RT}} + C$$

where ΔH_{sub} is the heat of sublimation and C is the integration constant.

4.10 Richard's Rule

A plot between heat of fusion at the melting point (ΔH_m) and melting temperature (T_m) for a number of metals has been outlined in Fig. 4.1. As shown in this figure, the value of (ΔH_m) varies linearly (approximately) with the melting point and the plot has a slope ($\Delta H_m / T_m$) of 8 – 17 J/K mol, i.e. the plot broadly obeys the following relationship:

$$\frac{\Delta H_m}{T_m} = \Delta S_m = 8 \sim 17\ \mathrm{J/K\,mol} \quad (4.101)$$

This correlation is known as Richard's rule and it states that the entropy change at the melting point (i.e. the ratio of latent heat of melting to the normal melting temperature) is constant for many of the solids. Equation (4.101) can be used to estimate the approximate value of ΔH_m at the melting point of the solid substance.

4.11 Trouton's Rule

A plot between heat of vaporization at the boiling point $\left(\Delta H_{\text{vap}}\right)$ and boiling temperature (T_b) for a number of metals has been shown in Fig. 4.2. As can be seen in this figure, the points are very close to a straight line and thus the value of ΔH_{vap} varies linearly (approximately) with the boiling point having a slope of 87.864 J/K mol, i.e. the plot broadly obeys the following relationship:

$$\frac{\Delta H_{\text{vap}}}{T_h} = \Delta S_{\text{vap}} = 87.864\ \mathrm{J/K\,mol} \quad (4.102)$$

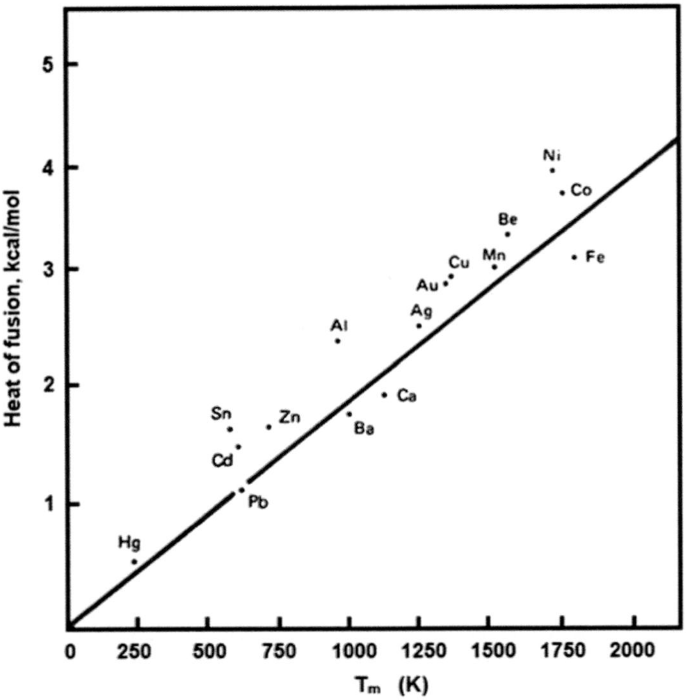

Fig. 4.1 Representation of Richard's rule

This correlation was first observed by Trouton and hence, was known as Trouton's rule. This relationship also indicates that the entropy of vaporization (ΔS_{vp}) of

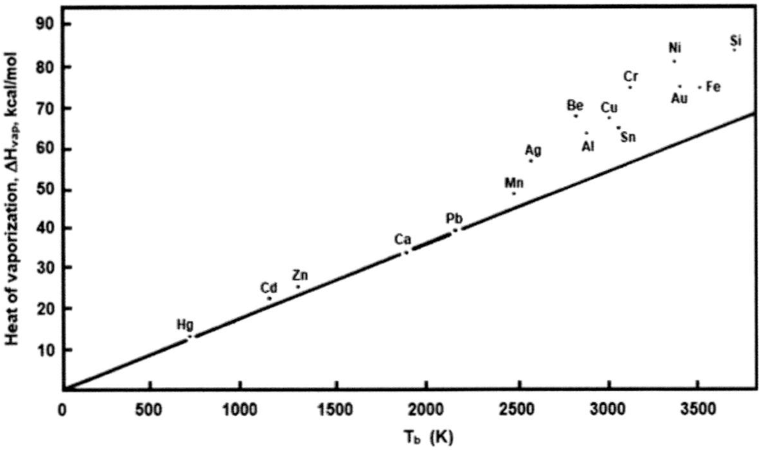

Fig. 4.2 Representation of Trouton's rule

any liquid (including both the metals and their compounds) at its boiling point is approximately same (87.864 J/K mol). Equation (4.102) appears to be very useful in estimation of the approximate value of ΔH_{vap} at the boiling point of the substance.

4.12 Third Law of Thermodynamics

In order to calculate changes in Gibbs and Helmholtz free energies for a process conducted at any temperature from the calorimetric data (i.e. ΔH, etc.), Gibbs–Helmholtz equations, as stated below in two alternate forms, can be used

$$\mathrm{d}\left(\frac{\Delta G}{T}\right) = -\frac{\Delta H}{T^2}\mathrm{d}T \text{ and } \mathrm{d}\left(\frac{\Delta A}{T}\right) = -\frac{\Delta U}{T^2}\mathrm{d}T$$

Integration of these equations leads to the following expressions:

$$\frac{\Delta G}{T} = -\int \frac{\Delta H}{T^2}\mathrm{d}T + I_P \text{ and } \frac{\Delta A}{T} = -\int \frac{\Delta U}{T^2}\mathrm{d}T + I_V$$

where I_P and I_V are the integration constants at constant pressure and volume conditions. For calculation of the values of ΔG and ΔA, a knowledge of these constants is required, but they cannot be determined either from 1st or 2nd laws of thermodynamics and use of empirical methods become necessary for their evaluations.

As outlined in Sect. 3.6.5, the variation of entropy with temperature at constant pressure and volume is expressed by the following equations:

At constant pressure, $\mathrm{d}S = C_P \mathrm{d}\ln T$ and at constant volume, $\mathrm{d}S = C_V \mathrm{d}\ln T$. Integration of these two equations within a temperature limit of 0 to T K yields the followings:

At constant pressure,

$$S_T = S_{0,P} + \int_0^T C_P \mathrm{d}\ln T \tag{4.103}$$

and at constant volume,

$$S_T = S_{0,V} + \int_0^T C_V \mathrm{d}\ln T \tag{4.104}$$

where $S_{0,P}$ and $S_{0,V}$ are the entropies at 0 K for processes occurring under constant pressure and volume conditions. Equations (4.91) and (4.92) indicate that for calculation of absolute value of entropy of a substance at any temperature T (i.e.S_T), the values of $S_{0,P}$ and must be known. 2nd law of thermodynamics gives only qualitative idea about the change in entropy and provides no information about the absolute value of entropy. Consideration of the numerical value of S_0 led to the enunciation of 3rd law of thermodynamics. Statement for the 3rd law of thermodynamics was formulated by W. Nernst in 1906 (known as Nernst's heat theorem), which was subsequently refined by Max Planck.

4.12.1 Nernst's Heat Theorem

In order to solve the problem involved in the calculation of absolute value of entropy at a temperature T, Nernst collected data on the enthalpies of different chemical reactions between pure solids or liquids, as obtained by calorimetric method, and determined their free energies as a function of temperature. These free energy and enthalpy data where plotted against temperature and yielded curves, as shown in Fig. 4.3.

It can be seen in this figure that the slopes of both the ΔG versus T and ΔH versus T curves tend to zero as temperature approaches 0 K. This led Nernst to arrive at the following conclusions:

$$\left[\frac{\partial(\Delta G^o)}{\partial T}\right]_P \to 0 \text{ as } T \to 0 \text{ and } \left[\frac{\partial(\Delta H^o)}{\partial T}\right]_P \to 0 \text{ as } T \to 0$$

From Eqs. (4.59) and (2.64), we know that $\left[\frac{\partial(\Delta G^o)}{\partial T}\right]_P = -\Delta S^o$ and $\left[\frac{\partial(\Delta H^o)}{\partial T}\right]_P = \Delta C_P^o$. Hence, one can write the above conclusions in the following alternative forms:

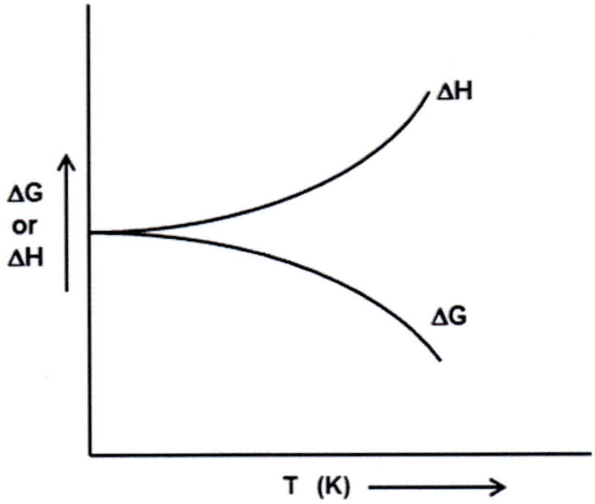

Fig. 4.3 Schematic diagram for the variation of ΔG and ΔH with temperature (T)

$$\Delta S^0 \rightarrow 0 \text{ and } \Delta C_P^0 \rightarrow 0 \text{ as } T \rightarrow 0$$

Nernst generalized these findings in the form of statement that for all reactions involving substances in condensed state, $\Delta S = 0$ at $T = 0\,K$. In other words, all isothermal processes at temperatures tending to 0 K occur without the variation in entropy. This is known as Nernst's heat theorem.

4.12.2 Latest Statement of Third Law of Thermodynamics

Let us consider the formation of a compound AB form elements A and B, i.e. $A(s) + B(s) = AB(s)$. The entropy change associated with this reaction will be as follows:

$$\Delta S^0 = S_{AB}^0 - S_A^0 - S_B^0 \tag{4.105}$$

As per Nernst's heat theorem, $\Delta S^0 = 0$ at $T = 0K$. Here, there are two possibilities: either (i) $S_{AB}^0 = S_A^0 + S_B^0$ or (ii) S_{AB}^0, S_A^0 and S_B^0 are all individually zero. The possibility of first alternative is very unlikely as a general feature. It may be true by chance in few cases. Hence, alternative (ii) is accepted as of general validity. This incompleteness of Nernst's statement was pointed out by Max Planck and the 3rd law of thermodynamics may now be enunciated as "the entropy of any pure and homogeneous substance (condensed state), which is in complete internal equilibrium, may be taken as zero at $T = 0$ K." Hence, the absolute value of entropy of a pure substance at any temperature can be determined by taking $T = 0$ K and $S_0 = 0$ as the lower limit of integration in Eqs. (4.103) and (4.104).

4.12.3 Concept of Complete Internal Equilibrium

Complete internal equilibrium in condensed state of the material means achievement of most ordered and stable structure with lowest energy state, which is obtained in the crystals when liquid is cooled under perfect equilibrium conditions. In actual practise, the rate of cooling from high temperature to 0 K is fast, i.e. far away from equilibrium cooling, and the atoms do not get sufficient time and energy to diffuse into the vacant sites and rearrange themselves in perfectly ordered manner. As a result, the more random atomic arrangement characteristics of high temperature are frozen in the crystals. In addition, these crystals consist of a large number of defects (e.g. vacancies, interstitials, dislocations, etc.) in their lattices and the unique most ordered atomic arrangement with stable structure is not obtained, i.e. the condensed phase has some entropy ($S \neq 0$) at 0 K. In case of formation of solutions also, the achievement of complete internal equilibrium becomes difficult due to randomness in the distribution of atoms/molecules of different substances and the entropy does not fall to zero at 0 K. Randomness in crystallographic orientation of molecules in the crystalline state can also give rise to non-zero entropy at 0 K.

The above considerations indicate that it is difficult to achieve the state of exactly zero entropy at $T = 0\,\text{K}$, which may cause errors in calculation of S_T values. However, for pure, crystalline and homogeneous substances, entropy at $T = 0\,\text{K}$ is very close to zero. So far as chemical reactions are concerned, the above phenomena are neglected and we always take $S = 0$ at $T = 0\,\text{K}$.

The requirement of complete internal equilibrium at 0 K in the statement of 3rd law of thermodynamics can be interpreted from statistical mechanical approach. If w represents the thermodynamic probability of an arrangement of atoms/molecules in the lattice, it can be related to the entropy by the relation $S = k \ln w$. Where k is a constant. From this equation, it is evident that any increase in the value of w in the material is accompanied by an increase in its entropy. When disorder in a substance completely vanishes, it becomes perfectly ordered and has only one possible arrangement of atoms (i.e $w = 1$). Then, a complete internal equilibrium is established in the material and entropy becomes zero at 0 K.

4.12.4 Verification of Third Law of Thermodynamics

The 3rd law can be verified in a number of ways. The quoted illustrations for the experimental verification of 3rd law of thermodynamics are mostly through studies on the allotropic transformations in different materials. In this section, the verification of this law is being considered through the following example of allotropic transformation:

Transformation of monoclinic sulphur into rhombic sulphur at 368.5 K: Monoclinic 'S' reversibly converts to rhombic 'S' at the transition temperature of 368.5 K. Let us calculate the entropy of monoclinic 'S' at $T_{tr} = 3685\,\text{K}$ by two different paths as follows:

(a) The monoclinic sulphur path—It consists of heating monoclinic 'S' from 0 to 368.5 K, i.e. $S(\text{monoclinic}, T = 0\,\text{K}) \rightarrow S(\text{monoclinic}, T = 368.5\,\text{K})$. The entropy change for this path, calculated by the following expression, has been reported to be 37.8 JK^{-1} mol^{-1}, i.e.

$$\Delta S_I = \int_0^{368.5} \frac{C_P(\text{momoclinic})}{T} dT = 37.8 \pm 0.4\,\text{JK}^{-1}\,\text{mol}^{-1}$$

where $\Delta S_I = S_{3685}(\text{monoclinc}) - S_0(\text{monoclinc})$. Hence, the absolute value of entropy of monoclinic 'S' at $T = 368.5\,\text{K}$ is given by the relation,

$$S_{368.5}(\text{monoclinc}) = 37.8 + S_0(\text{monoclinc}) \quad (4.106)$$

where S_0(monoclinc) is the absolute value of monoclinic ‘S’ at 0 K.

(b) The rhombic sulphur path—It consists of heating rhombic ‘S’ from 0 to 368.5 K and transformation of this form of ‘S’ to monoclinic ‘S’ at this temperature. It can be represented as:

$$\begin{aligned} S(\text{rhombic at } T = 0\,\text{K}) &\rightarrow S(\text{rhombic at } T = 368.5\,\text{K}) \\ &\rightarrow S(\text{monoclinic at } T = 368.5\,\text{K}) \end{aligned}$$

The net entropy change for path (ii) has been calculated as follows:

$$\Delta S_{\text{II}} = \int_0^{368.5} \frac{C_P(\text{rhombic})}{T}\text{d}T + \frac{\Delta H_{\text{tr}}(\text{rhombic} \rightarrow \text{monoclinic})}{368.5}$$

$$= 36.8 + 1.09 = 37.9 \pm 0.2\,\text{JK}^{-1}\,\text{mol}^{-1}$$

where $\Delta S_{\text{II}} = S_{3685}(\text{monoclinc}) - S_0(\text{rhombic})$. Therefore, the absolute value of entropy of monoclinic ‘S’ at $T = 368.5$ K can be expressed as:

$$S_{3685}(\text{monoclinc}) = 37.9 + S_0(\text{rhombic}) \tag{4.107}$$

where S_0 (rhombic) is the absolute value of entropy of rhombic ‘S’ at 0 K. Comparison of Eqs. (4.94) and (4.95) clearly indicates that in both the cases, the absolute value of entropy of monoclinic ‘S’ at 368.5 K is the same within the limits of experimental error under the assumption that both the monoclinic and rhombic sulphur have zero entropy at 0 K. Hence, the 3rd law is verified.

Similar experimental studies on other materials, such as tin, phosphine, have supported the statement of 3rd law.

4.12.5 Consequences of Third Law of Thermodynamics

Some very important effects of the 3rd law, leading towards the determination of few thermodynamic properties, are as follows:

(i) Thermal coefficients of volumetric expansion and pressure at 0 K: The first consequence of the 3rd law states that the thermal coefficients of volumetric expansion (α) and pressure (η) become equal to zero as temperature approaches 0 K. This statement can be proved as follows:

From Eq. (4.88), we know that $\alpha = \frac{1}{V}\left(\frac{\delta V}{\delta T}\right)_P$.
According to Maxwell's relation outlined in Eq. (4.77), $\left(\frac{\delta S}{\delta P}\right)_T = -\left(\frac{\delta V}{\delta T}\right)_P$.
Hence,

$$\alpha = \frac{1}{V}\left(\frac{\delta V}{\delta T}\right)_P = -\frac{1}{V}\left(\frac{\delta S}{\delta P}\right)_T \tag{4.108}$$

The 3rd law postulates that $S \to 0$ (i.e. a constant value) as $T \to 0$ K. Hence,

$$\left(\frac{\delta S}{\delta P}\right)_T = 0 \text{ as } T \to 0\,\mathrm{K}$$

Now, Eq. (4.108) becomes as $\alpha = \frac{1}{V}\left(\frac{\delta V}{\delta T}\right)_P = -\frac{1}{V}\left(\frac{\delta S}{\delta P}\right)_T = 0$ as $T \to 0\,\mathrm{K}$
The thermal coefficient of pressure (η) is defined by the following expression:

$$\eta = \frac{1}{P}\left(\frac{\delta P}{\delta T}\right)_V \tag{4.109}$$

From Maxwell's relation outlined in Eq. (4.76), we know that $\left(\frac{\delta S}{\delta V}\right)_T = \left(\frac{\delta P}{\delta T}\right)_V$
Hence, Eq. (4.109) becomes

$$\eta = \frac{1}{P}\left(\frac{\delta P}{\delta T}\right)_V = \frac{1}{P}\left(\frac{\delta S}{\delta V}\right)_T \tag{4.110}$$

As stated above, $S \to 0$ (i.e. a constant value) as $T \to 0\,\mathrm{K}$. Hence, $\left(\frac{\delta S}{\delta V}\right)_T = 0$ as $T \to 0\,\mathrm{K}$

Now, Eq. (4.110) becomes $\eta = \frac{1}{P}\left(\frac{\delta S}{\delta V}\right)_T = 0$ as $T \to 0\,\mathrm{K}$.

(ii) Calculation of absolute value of entropy: As outlined in Eq. (4.104), the entropy of a substance at any temperature T under constant volume condition is given by the expression,

$$S_T = S_{0,V} + \int_0^T C_V \,\mathrm{d}\ln T$$

According to 3rd law, $S_{0,V} = 0$ at $T = 0$ K. Hence, $S_T = \int_0^T C_V \,\mathrm{d}\ln T$

The value of C_V can be determined experimentally by calorimetric technique. Hence, one can calculate the absolute value of entropy of a substance at any temperature T (i.e. S_T) by using the above formula. Similarly, the value of S_T under isobaric condition can be determined by replacing the term C_V by C_p in the above formula.

(iii) Calculation of free energy change for a process: 3rd law also helps in calculation of free energy change during a chemical reaction by the use of calorimetric data obtained in the form of enthalpy as a function of temperature. We know that $\Delta G = \Delta H - T\Delta S$. The change in entropy of the reaction at any temperature T is expressed by the following:

$$\Delta S = \int_0^T \frac{\Delta C_P}{T} \mathrm{d}T$$

As illustrated in Chap. 3, one can write the following expression for ΔH of a reaction:

$$\Delta H = \Delta H_0 + \int_0^T \Delta C_P \mathrm{d}T$$

where ΔH_0 is the change in enthalpy at 0 K. with the help of these expressions of ΔS and ΔH, one can rewrite the equation $\Delta G = \Delta H - T\Delta S$ in the following form:

$$\Delta G = \Delta H_0 + \int_0^T \Delta C_P \mathrm{d}T - T\int_0^T \frac{\Delta C_P}{T} \mathrm{d}T \tag{4.111}$$

All the quantities involved in this expression can either be determined by calorimetric method or calculated by using Debye's theory of specific heat. Thus, one can calculate free energy change of a reaction at any desired temperature.

(iv) By definition, Gibbs free energy $G = H - TS$ and Helmholtz free energy $A = U - TS$. From 3rd law of thermodynamics, $S = 0$ at $T = 0\,\mathrm{K}$. Hence, $G_0 = H_0$ and $A_0 = U_0$.
(v) From Boltzmann equation, we know that $S = K \ln w$ and the 3rd law states that $S = 0$ at $T = 0$ K. Thus, $\ln w = 0$ or $w = 1$. Hence, for 3rd law to be applicable, the material must be in perfect ordered condition.

4.12.6 Significance of Third Law of Thermodynamics

Second law allows only the measurement of entropy changes in thermodynamic processes and the absolute value of entropy is not determinable from this law. Third law defines entropy of substances at absolute zero and thereby outlines a method to calculate the absolute values of entropy of substances at any desired temperature, which are ultimately required to find out energy changes during thermodynamic processes. Both the 1st and 2nd laws led to the development of new state functions, namely the internal energy and entropy, but 3rd law does not introduce any new state function and it only provides help in calculation of the absolute value of entropy. In addition, there exists no theoretical prove for 1st and 2nd laws, but the 3rd law can be derived from quantum statistics.

Solved Problems

Problem 4.1 Find out the values of ΔH^o and ΔS^o for the reaction $2Cu(s) + 0.5O_2(g) = Cu_2O(s)$ at 25 °C.. Data given: ΔG^o for this reaction = – 40,538.76 – 3.92 $T\log T$ + 29.53 T cal/mol.

Solution
From Gibbs–Helmholtz equation, we know that $\frac{d(\Delta G^o/T)}{dT} = -\frac{\Delta H^o}{T^2}$.
Now,

$$\Delta G^o/T = -40{,}538.76/T - 3.92\log T + 29.53$$
$$= -40{,}538.76/T - 1.702\ln T + 29.53$$

$$\frac{d(\Delta G^o/T)}{dT} = 40{,}538.76/T^2 - 1.702/T = -\frac{\Delta H^o}{T^2}$$

or $\Delta H^o = -40{,}538.76 + 1.702T$

Now, the value of ΔH^o at 298 K $= -40{,}538.76 + 1.702 \times 298 = -40{,}031.56$ cal/mol.

The value of ΔG^o at 298 K $= -40{,}538.76 - 3.92 \times 298 \times \log 298 + 29.53 \times 298$. $= -34{,}629.1$ cal/mol $= -34.629$ kcal/mol.

We know that $\Delta G^o = \Delta H^o - T\Delta S^o$ or $\Delta S^o = \frac{\Delta H^o - \Delta G^o}{T}$ or $\Delta S^o = (-40.032 + 34.629)/298 = -0.018$ kcal/K mol $= -75.78$ J/K mol.

Problem 4.2 By calculation, determine the normal boiling point of titanium (Ti) with the help of the following data:

Vapour pressure of Ti (l) at 2500 K = 200 N/m^2

Heat of vaporization (ΔH_{vap}) at the normal boiling point of Ti = 435.14 kJ/mol.

Solution
Let the normal boiling point of Ti is T_b at 1 atmospheric pressure (i.e. 101,325 N/m^2).

From Clausius–Clapeyron equation, we know that $\ln P_2 - \ln P_1 = -\frac{\Delta H_{vap}}{R}\left(\frac{1}{T_2} - \frac{1}{T_1}\right)$.

Now, $T_1 = 2500$ K, $P_1 = 200$ N/m^2, $T_2 = Tb$ and $P_2 = 101{,}325$ N/m^2.

On inserting these values in the above equation, we get the following:

$$\ln(101{,}325/200) = -\left(435.14 \times 10^3/8.314\right)[1/T_b - 1/2500]$$

or $6.23 = -52{,}338.22\,(1/T_b - 0.0004)$

or $-1.19 \times 10^{-4} + 0.0004 = 1/T_b$.

or $T_b = 3559$ K.

Problem 4.3 By applying Trouton's equation, find out the vapour pressure of zinc (Zn) at 1073 K. The normal boiling point of Zn is 1180 K.

Solution

From Trouton's equation, we know that $\frac{\Delta H_{\text{vap}}}{T_b} = 87.864$ J/K mol.

Now, $\frac{\Delta H_{\text{vap}}}{1180} = 87.864$ or $\Delta H_{\text{vap}} = 103{,}679.52$ J/mol.

From Clausius–Clapeyron equation, we know that $\ln P_2 - \ln P_1 = -\frac{\Delta H_{\text{vap}}}{R}\left(\frac{1}{T_2} - \frac{1}{T_1}\right)$.

Now, $T_1 = 1073$ K, $T_2 = 1180$ K, $P_1 =?$ and $P_2 = 1$ atm.

On inserting these values in the above equation, we get the following:

$$\ln(1/P_1) = -103{,}679.52/8.314[1/1180 - 1/1073]$$

$$\text{or} \quad \ln(1/P_1) = 1.054$$

or $1/P_1 = 2.869$ or $P_1 = 0.35$ atm.

Problem 4.4 α-Fe transforms into γ-Fe at a temperature of 1183 K and 1 atm pressure. Find out the change in this equilibrium transformation temperature (T_{tr}) on increasing the pressure from 1 to 100 atm. Data given are as follows:

Heat of the above transformation $\left(\Delta H_{\alpha\rightarrow\gamma}\right) = 218.18$ cal/mol.

Densities of α-Fe and γ-Fe = 7.571 and 7.633 g/cm^3

Solution

We know that $\left(\frac{dP}{dT}\right)_{\text{eq}} = \frac{\Delta H_{\text{tr}}}{T\Delta V_{\text{tr}}}$ or $\left(\frac{dT}{dP}\right)_{\text{eq}} = \frac{T_{\text{tr}}\Delta V_{\text{tr}}}{\Delta H_{\text{tr}}}$.

Now, $T_{\text{tr}} = 1183$ K, $\Delta H_t = 218.18$ cal/mol and atomic weight of Fe = 55.85 g.

Hence, volume of α-Fe = 55.85/7.571 = 7.377 cm^3

Volume of γ-Fe = 55.85/7.633 = 7.317 cm^3

Volume change during transformation from x-Fe to γ-Fe(ΔV_{tr}) = 7.317 − 7.377 = −0.06 cm^3.

As is known to us, 1 cal = 41.193 cm^3.atm.

Now, $\frac{dT}{dP} = \frac{-0.06\times 1183}{218.18\times 41.193} = -0.00789$ K/atm.

When $P = 1$ atm then $T_{\text{tr}} = 1183$ K.

When $P = 100$ atm then $dP = 100 - 1 = 99$ atm.

Hence, $dT = -99 \times 0.00789 = -0.782$ K, i.e. change in transformation temperature (ΔT_{tr}) = −0.782 K.

This result shows that the transformation temperature (T_{tr}) decreases by a negligible amount on increasing pressure from 1 to 100 atm.

Problem 4.5 Solid lead melts at a temperature of 600 K and 1 atm pressure. The heat of melting of lead at its normal melting point has been found to be 4974 J/mol. Prove that the melting point of lead practically remains the same on increasing the pressure from 1 to 100 atm. Data given are as follows:

Densities of solid and liquid lead at its melting point = 11.05 and 10.645 g/cm^3.

Solution

Atomic weight of lead (Pb) = 207 g, $\Delta H_m = 4974$ J/mol = 1190 cal/mol, $T_m =$ 600 K.

Now, volume of solid Pb = 207/11.05 = 18.733 cm^3

Volume of liquid Pb = 207/10.645 = 19.446 cm^3

Volume change during transformation from solid Pb to liquid Pb(ΔV_m) = 19.446 − 18.733 = 0.713 cm^3.

We know that $\left(\frac{dT}{dP}\right)_{eq} = \frac{T_m \Delta V_m}{\Delta H_m}$.

As is known to us, 1 cal = 41.193 cm^3.atm.

Now, $\frac{dT}{dP} = \frac{0.713 \times 600}{1190 \times 41.193} = 0.00873$ K/atm.

When $P = 1$ atm then $T_m = 600$ K.

When $P = 100$ atm then dP = 100 − 1 = 99 atm.

Hence, d$T = \Delta T_m = 99 \times 0.00873 = 0.864$ K.

The result indicates that the melting point of Pb increases by only 0.864 K on increasing pressure from 1 to 100 atm. This value is negligible and hence, melting point practically remains unchanged.

Problem 4.6 Graphite and diamond are the two allotropes of carbon and graphite is the stable form at a temperature of 298 K and pressure of 1 atm. Find out the value of minimum pressure required which needs to be applied for conversion of graphite to diamond at a temperature of 298 K. Data given are as follows:

Densities of graphite and diamond at 298 K = 2.22 and 3.515 g/cm^3

Entropy (S^o) at 298K for graphite and diamond = 1.37 and 0.58 cal/K mol.

Heat of transformation (ΔH^o) at 298 K = H^o_{no}(diamond) − H^o_{no}(graphite) = 454.55 cal/mol.

Solution

Thermodynamically, transformation from graphite to diamond will be feasible when ΔG(graphite → diamond) is negative (i.e. < 0).

We know that $dG = VdP - SdT$.

At constant temperature,

$$dG = VdP \quad \text{or} \quad d(\Delta G) = \Delta V \cdot dP \tag{4.112}$$

where ΔV = Change in volume during transformation from graphite to diamond

Now, Volume of graphite = 12/2.22 = 5.405 cm^3

Volume of diamond = 12/3.515 = 3.414 cm^3

Hence, $\Delta V_{tr} = 3.414 - 5.405 = -1.991$ cm^3.

At a temperature of 298 K and 1 atm pressure,

$$\Delta G^\circ(\text{graphite} \rightarrow \text{diamond}) = \Delta H^\circ(\text{graphite} \rightarrow \text{diamond}) - T\Delta S^\circ(\text{graphite} \rightarrow \text{diamond})$$

or ΔG° (graphite → diamond) = 454.55 − 298 × (0.58 − 1.37) = 689.97 cal/mol = 689.97 × 41.193 = 28,421.93 cm^3.atm/mol.

(Because, 1 cal = 41.193 cm^3.atm).

At the minimum pressure required for transformation, ΔG^{o}(graphite → diamond) = 0.

Integration of Eq. (4.112) yields, $\int_{28,421.93}^{0} d(\Delta G) = \int_{1}^{P_{min}} \Delta V \cdot dP$ or $[\Delta G]_{28,421.93}^{0} = -1.991 \times [P]_{1}^{P_{min}}$ or $28{,}421.93 = 1.991(P_{min} - 1)$ or $P_{min} = 14{,}276.21$ atm.

Problem 4.7 Following is the formula for the vapour pressure of liquid Zn above its melting point of 419 °C:

$$\ln p_{Zn}^{o}(\text{vap}) = \frac{-14.765 \times 10^3}{T} + 12.805$$

By derivation, establish a formula for the vapour pressure over solid Zn. Data given is as follows:

Heat of melting (ΔH_m) = 1596.41 cal/mol.

Solution

Melting point of pure Zn = 419 °C = 692 K.

At this temperature of 692 K, solid Zn is in equilibrium with liquid Zn.

Hence, at 692 K, p_{Zn}^{o}(vap)(over solid Zn) = p_{Zn}^{o}(vap)(over liquid Zn).

From the equation provided in the question, in

$$p_{zn}^{o}(vap) = \frac{-14.765 \times 10^3}{692} + 12.805 = -8.53 \tag{4.113}$$

From Clausius–Clapeyron equation, we know that

$$\ln p_{\text{zn}}^{o}(\text{vap}) = \frac{-\Delta H^{o}(\text{vap})}{\text{RT}} + C \tag{4.114}$$

$$\ln p_{\text{Zn}}^{o}(\text{sub}) = \frac{-\Delta H^{o}(\text{sub})}{\text{RT}} + C' \tag{4.115}$$

where C and C' are integration constants.

Now, heat of sublimation ΔH^{o}(sub) = ΔH^{o}(vap) + ΔH^{o}(melt).

where ΔH^{o}(vap) and ΔH^{o}(melt) are heats of vaporization and melting.

On comparing Eq. (4.114) with the equation provided in the question, we get

$$\frac{-14.765 \times 10^3}{T} + 12.805 = \frac{-\Delta H^{o}(\text{vap})}{\text{RT}} + C$$

$$\frac{-\Delta H^{o}(\text{vap})}{\text{RT}} = \frac{-14.765 \times 10^3}{T} \text{ and } C = 12.805$$

or $V^o(\text{vap}) = 14.765 \times 10^3 \times R = 14.765 \times 10^3 \times 1.98 = 29{,}234.7$ cal/mol

Hence, $\Delta H^o(\text{ sub }) = 29{,}234.7 + 1596.41 = 30{,}831.11$cal/mol.

Now, on inserting this value in Eq. (4.115), we get $\ln p^o_{zn}(sub) = \frac{-30{,}831.11}{1.98\times 692} + C' = -22.5 + C'$.

Since, $\ln p^o_{Z_n}(\text{sub}) = \ln p^o_{Z_n}(\text{vap})$ or $-22.5 + C' = -8.53$ or $C' = 13.97$.

Hence, the required equation is: $\ln p^o_{Z_n}(\text{sub}) = \frac{-30{,}831.11}{1.98\times T} + 13.97 = \frac{-15.571\times 10^3}{T} + 13.97$.

Problem 4.8 α-Fe and δ-Fe (both BCC) and γ-Fe (FCC) are the allotropes of Fe. γ-Fe transforms into δ-Fe at a temperature of 1673 K and heat of transformation ($\Delta H_{\gamma\to\delta}$) at this temperature is 877.8 J/mol. The values of heat capacity of these phases are as follows:

C_P for γ-Fe = $7.69 + 19.48 \times 10^{-3}$ T J/K mol.

C_P for δ-Fe = 43.89 J/K mol.

Derive necessary expressions for ΔH and ΔG in terms of temperature for transformation from γ-Fe to δ-Fe.

Solution

Calculation for ΔG for transformation from γ-Fe to δ-Fe:

$\Delta C_P = C_P(\delta\text{-Fe}) - C_P(\gamma\text{-Fe}) = 43.89 - 7.69 - 19.48 \times 10^{-3}\ T = 36.2 - 19.48 \times 10^{-3}\ T$.

We know that $\Delta C_P = \mathrm{d}\Delta H/\mathrm{d}T$. Hence, $\mathrm{d}\Delta H = (36.2 - 19.48 \times 10^{-3}\ T)\mathrm{d}T$.

On integration of this equation, we get

$$\Delta H = 36.2T - 9.74 \times 10^{-3}T^2 + C \tag{4.116}$$

where C is the integration constant.

At 1673 K, ΔH = 877.8 J/mol. Thus, $877.8 = 36.2 \times 1673 - 9.74 \times 10^{-3} \times 1673^2 + C$.

or $C = -32{,}423.23$.

Now, Eq. (4.116) can be written as $\Delta H = 36.2\,T - 9.74 \times 10^{-3}T^2 - 32{,}423.23$ J/mol.

This is the required expression of ΔH in terms of T for transformation from γ-Fe to δ-Fe.

Gibbs–Helmholtz equation yields

$$\frac{\mathrm{d}(\Delta G/T)}{\mathrm{d}T} = -\frac{\Delta H}{T^2}$$

or

$$\frac{\mathrm{d}(\Delta G/T)}{\mathrm{d}T} = -\frac{\Delta H}{T^2} = -\frac{36.2}{T} + \frac{32{,}423.23}{T^2} + 9.74 \times 10^{-3}$$

or

$$d(\Delta G/T) = \left(-\frac{36.2}{T} + \frac{32{,}423.23}{T^2} + 9.74 \times 10^{-3}\right)dT.$$

On integration, we get the following: $\frac{\Delta G}{T} = -36.2 \ln T - \frac{32423.23}{T} + 9.74 \times 10^{-3}T + C'$ or

$$\Delta G = -36.2T \ln T - 32423.23 + 9.74 \times 10^{-3}T^2 + C'T \tag{4.117}$$

where C' is the integration constant.

At a temperature of 1673 K, γ-Fe and δ-Fe phases are in equilibrium. Hence, $\Delta G_{\gamma \to \delta} = 0$.

Now, $C' = 36.2 \times \ln 1673 + (32{,}423.23/1673) - 9.74 \times 10^{-3} \times 1673 = 271.77$.

Now, Eq. (4.117) becomes as $\Delta G = -36.2T \ln T - 32423.23 + 9.74 \times 10^{-3}T^2 + 271.77T$.

This is the required expression of ΔG in terms of T for transformation from γ-Fe to δ-Fe.

Exercise

4.1 By using Clausius–Clapeyron equation, find out the value of vapour pressure of liquid Cu at 2273 K. Following data are given:

Boiling point of Cu = 2868 K, Heat of vaporization = 73.07 kcal/mol.

[**Ans**: 0.035 atm].

4.2 Find out the value of minimum pressure required to stabilize cementite (Fe_3C) at a temperature of 25 °C. Data given are as follows:

ΔG^o[for the reaction 3Fe (s) + C (s) = Fe_3C (s)] = 4556.22 cal/mol at 25 °C.

Molar volumes of Fe, Carbon and Fe_3C = 7.10, 5.34 and 24.25 cm^3

[**Ans**: 78,450 atm].

4.3 Using Trouton's rule, find out the value of heat of vaporization of liquid silver (Ag) at its normal boiling point of 1874 °C. Also find out the value of difference between heat capacities of liquid and gaseous Ag.

[**Ans**: 260.6 kJ/mol; 7.07 J/K mol].

4.4 At a Temperature of 1809 K, δ-Fe Melts with a Latent Heat of Melting of 13,794 J/mol. The Values of Heat Capacity of These Phases Are as Follows:

C_P for δ-Fe = 43.89 J/K mol.

C_P for Fe (l) = 41.8 J/K mol.

Derive necessary expressions for ΔH and ΔG in terms of temperature for transformation from δ-Fe to liquid Fe.

[**Ans**: ΔH = 17,577 – 2.09 T J/mol; ΔG = – 25.41 T + 17,577 + 2.09Tln$T$ J/mol].

Chapter 5
Fugacity, Activity, Equilibrium Constant and Ellingham Diagram

This chapter has focused on the potential thermodynamic topics, e.g. activity, fugacity, equilibrium constant and Ellingham diagrams, and the entire discussion has been covered in seven sections. An evaluation of equilibrium constant is helpful in understanding the equilibrium state of a chemical reaction or the desired direction in which the reaction has to be moved. ΔG^0 (standard free energy change) versus T plots in the form of Ellingham diagram provides a fundamental knowledge in clear-cut understanding of oxidation and reduction reactions which are essential to have a knowledge on smelting and refining processes.

Fugacity along with its variation with temperature and pressure, thermodynamic activity and concept of equilibrium constant and its dependence on temperature and pressure have been discussed in Sects. 5.1–5.3. Some powerful and extensively used equations, e.g. $\Delta G^0 = -RT \ln K$, $\Delta G = RT \ln a$, $\Delta G = RT \ln f$, etc. have been explained adequately in this chapter. In addition to these, this chapter provides a complete and easily understandable interpretation of Ellingham diagrams in terms of their construction, relative stabilities of metal oxides and sulphides, phase transformation, carbothermic and metallothermic reductions of metal oxides, dissociation of metal oxides, deoxidation of steel, etc. Salient features and limitations of an Ellingham diagram have also been covered in Sects. 5.6 and 5.7. This chapter has ended with sufficient number of solved and unsolved problems.

5.1 Fugacity

From Eq. (4.49), we know that $\mathrm{d}G = RT\mathrm{d}\ln P$. On integration, it leads to $G = G^0 + RT \ln P$. The derivation of this equation is based on the ideal gas law (i.e. $PV = RT$) and is, therefore, applicable only to ideal gases or their mixtures. This equation shows that at any temperature, the free energy of an ideal gas is a linear function of the logarithm of its pressure. When a gas deviates from ideality, the

S. K. Sahoo et al., *Fundamentals of Metallurgical Thermodynamics*,
https://doi.org/10.1007/978-981-99-6671-4_5

relation between free energy and logarithm of its pressure (i.e. $\ln P$) does not remain linear and poses some difficulties in calculation of some thermodynamic properties, such as ΔG. In order to overcome this problem, Lewis introduced a new function called fugacity (f) in place of P in Eq. (4.49), and this ensured linearity between G and lnf. Thus, fugacity is mathematically related to the free energy by the following expression:

$$dG = RTd\ln f \tag{5.1}$$

Here, we have two cases as follows:

Case I: On equating the above two equations of dG, we find that for an ideal gas, $f = P$ at all pressures. Hence, the equation $dG = RTd\ln P$ is applicable.

Case II: For non-ideal or real gases, fugacity is not equal to pressure, i.e. $f \neq P$. Further, all gases tend to exhibit ideal behaviour at very low pressures and/or high temperatures. This can be expressed as follows:

$$\text{As } P \to 0,\ f \to P \text{ or as } P \to 0,\ f/P \to 1$$

Thus, for non-ideal gases, $f = P$ only at extremely low pressures, i.e. the expression $\mathrm{d}G = RTd\ln P$ is applicable only at low pressures. The state of a pure gas at one atmospheric pressure and the temperature of interest is referred to as its standard state. In terms of fugacity, the standard state for any of the gas now is defined as the state in which $f = 1$ at temperature T.

Let us consider a gas, which obeys the equation of state as follows:

$$V = \frac{RT}{P} - \alpha \tag{5.2}$$

where α is a function of temperature and is a measure of the deviation of the gas from ideality. Equation (4.22) gives $\mathrm{d}G = V\mathrm{d}P$ at constant temperature T, and Eq. (5.1) gives $\mathrm{d}G = RTd\ln f$ at constant temperature T.

Thus,

$$V\mathrm{d}P = RTd\ln f$$

or

$$\left(\frac{RT}{P} - \alpha\right)\mathrm{d}P = RTd\ln f \quad \text{or} \quad RT\frac{\mathrm{d}P}{P} - \alpha\mathrm{d}P = RTd\ln f$$

or

$$RTd\ln f - RTd\ln P = -\alpha\mathrm{d}P$$

or

$$d \ln\left(\frac{f}{P}\right) = -\frac{\alpha}{RT} dP \tag{5.3}$$

Integration between the pressure limits 0 to P at constant temperature yields the following:

$$\ln\left(\frac{f}{P}\right)_P - \ln\left(\frac{f}{P}\right)_0 = -\frac{\alpha P}{RT}$$

$$\text{At } P = 0,\ f/P = 1 \text{ and thus } \ln\left(\frac{f}{P}\right)_0 = 0$$

Hence,

$$\ln\left(\frac{f}{P}\right) = -\frac{\alpha P}{RT} \quad \text{or} \quad \frac{f}{P} = e^{-\frac{\alpha P}{RT}}$$

We also know that

$$e^{-\frac{\alpha P}{RT}} = 1 - \frac{\alpha P}{RT}$$

Hence,

$$\frac{f}{P} = 1 - \frac{\alpha P}{RT} = 1 - \left(\frac{RT}{P} - V\right)\frac{P}{RT} = \frac{PV}{RT}$$

If the gas behaves ideally, then

$$P_{\text{ideal}} = \frac{RT}{V}$$

Thus,

$$\frac{f}{P} = \frac{P}{P_{\text{ideal}}}$$

This indicates that for small deviations from ideality, the actual pressure (P) exerted by the gas is the geometric mean of the fugacity and ideal pressure.

5.1.1 Physical Significance of Fugacity

As discussed in preceding chapter, a chemical reaction or physical change proceeds only in a direction in which free energy of the system decreases. Application of this statement in equation $\Delta G = RT \ln f$ clarifies that a component undergoing a change in any system will tend to move towards a state of lower fugacity. In other words, it can be stated that a component under given set of external conditions will tend to escape to a new state or other part of the system or even to the surroundings where its fugacity will be lower than that in the initial state. High gas pressure indicates the tendency of the gas molecules to escape outside the container. Similarly, high fugacity indicates a greater tendency of a component or substance to escape, dissolve, intermix or react. Therefore, property of fugacity is very similar to pressure which determines the state of mechanical equilibrium in any system. On the basis of this argument, fugacity may be defined as a measure of the escaping tendency of a component in a system.

In the extraction of metals and their refining processes, we are usually concerned with gases at low pressures and relatively high temperatures, where deviations from ideal behaviour of gases are usually not significant. Under such conditions, replacement of the term f (fugacity) by the actual partial pressure (P) of the gas is well within the experimental uncertainty limits. But at high pressures and low temperatures, the deviation from ideal behaviour is often pronounced and disregard of the fugacity may give rise to large errors.

5.1.2 Fugacity of Solids and Liquids

In comparison with the fugacity of gas, the fugacity values of solids and liquids are very small, i.e. $f_g > f_l > f_s$. The fugacity values of solids and liquids can be taken equal to their vapour pressures. These materials have low vapour pressures and, hence, can be equated to their fugacity values. The condensed materials (i.e. solids or liquids) having lower boiling points have higher escaping tendencies or fugacity and vice versa. Let us assume a condensed phase to be in equilibrium with its vapour phase of the same free energy (i.e. $G_{\text{condensed}} = G_{\text{vapour}}$). As per discussion made in Sect. 5.1, fugacity of the vapour phase will be equal to that of the condensed phase because these phases are in equilibrium with each other.

5.1.3 Variation of Fugacity with Pressure

The fugacity can be dealt with in terms of compressibility factor Z. From Eq. (5.3), we have

$$d\ln\left(\frac{f}{P}\right) = -\frac{\alpha}{RT}\mathrm{d}P = -\left(\frac{RT - PV}{PRT}\right)\mathrm{d}P = \left(\frac{PV}{RT} - 1\right)\frac{\mathrm{d}P}{P}$$

We know that $Z = PV/RT$. Hence, the above equation becomes as follows:

$$d\ln\left(\frac{f}{P}\right) = \left(\frac{Z-1}{P}\right)\mathrm{d}P$$

Let the state of one mole of gas change from (P_1, T) to (P_2, T). On integration between these limits, we have the following:

At constant temperature

$$\left[\ln\left(\frac{f}{P}\right)\right]_{P_1}^{P_2} = \int_{P_1}^{P_2} \frac{Z-1}{P}\mathrm{d}P \tag{5.4}$$

Fugacity may be evaluated by direct integration if Z is known as a function of P. Let us suppose that $Z = 1 + BP + CP^2 + \ldots$ where B, C, etc. are constants. Now, Eq. (5.4) becomes as

$$\left[\ln\left(\frac{f}{P}\right)\right]_{at\,P_2} - \left[\ln\left(\frac{f}{P}\right)\right]_{at\,P_1} = \int_{P_1}^{P_2} \left(\frac{BP + CP^2 + \ldots}{P}\right)\mathrm{d}P$$

$$= B(P_2 - P_1) + \frac{C}{2}(P_2^2 - P_1^2) + \ldots \tag{5.5}$$

This is the required equation showing the variation of f with P of a gas.

5.1.4 Variation of Fugacity with Temperature

Let G and G^* are free energies of the gas at pressures P and P^* and temperature T. Integration of Eq. (5.1) yields the following:

$$G = RT\ln f + G^0 \tag{5.6}$$

If P^* is so small that $f = P^*$, then

$$G^* = RT\ln P^* + G^0 \tag{5.7}$$

Subtraction of Eq. (5.7) from Eq. (5.6) yields the following:

$$\frac{G}{T} - \frac{G^*}{T} = R\ln f - R\ln P^* = R\ln\left(\frac{f}{P^*}\right)$$

Differentiation of this equation at constant pressure with respect to $1/T$ gives the following relationship:

$$R\frac{d\ln f}{d(1/T)} = \frac{d(G/T)}{d(1/T)} - \frac{d(G^*/T)}{d(1/T)}$$

On substituting values from Eq. (4.67), one gets the following relation:

$$R\frac{d\ln f}{d(1/T)} = H_P - H^*_{P*} \tag{5.8}$$

where H_P and H^*_{P*} are the enthalpies of the gas at temperature T and pressures P and P^*. Equation (5.8) forms the required relationship for the variation of fugacity with temperature. In order to make the use of this form of equation easier, it is advisable to express the right-hand side of the equation in terms of easily determinable experimental parameters.

5.2 Thermodynamic Activity

In chemical thermodynamics, activity is a measure of (i) an effective concentration of the material to be available for a chemical reaction, (ii) the capacity of a substance to undergo a chemical change and (iii) the ability of a substance to take part in a chemical reaction. It is known to us that the thermodynamic properties, such as H, S and G, of a substance in pure state are different from those in the state of solution, i.e. when a pure element goes into the solution, the state of aggregation/activity change and the thermodynamic properties also alter. This means that apart from temperature and pressure, change in composition affects the thermodynamic properties of materials. The concept of activity offers an index to measure the changes in thermodynamic properties when a pure element goes into the solution. The activity of a substance depends upon all the factors that alter its thermodynamic properties. These include temperature, pressure, concentration/composition, nature of groupings between atoms or molecules, standard state chosen, presence of an electric or magnetic field, etc. Activity and its coefficient in context to ideal and non-ideal solutions have also been discussed in detail in the next coming chapter.

Thermodynamic activity of a component (a) is defined as the ratio of its fugacity in the existing (i.e. actual) state to that in the chosen standard state and is a dimensionless quantity. Mathematically, it is expressed as follows:

$$a = \frac{f}{f^0} \quad \text{Or} \quad f = f^0 a \tag{5.9}$$

For i - th ih component in a solution,

$$a_i = \frac{f_i}{f_i^0} \quad \text{Or} \quad f_i = f_i^0 a_i \tag{5.10}$$

where f_i = fugacity of the component i in the existing state/solution at temperature T and f_i^0 = fugacity of species i in the standard state (i.e. pure form) at the same temperature. For a pure substance, $f_i = f_i^0$ and $a_i = 1$, i.e. the activities of pure substances are unity. The choice of standard state is arbitrary, and the value of activity depends upon it. However, there are certain conventional standard states which are usually used in different solutions, as outlined in Chap. 6.

Now, Eq. (5.1) becomes as follows:

$$\mathrm{d}G = RTd\ln f = RTd\ln\left(f^0 a\right) = RT\left(d\ln f^0 + d\ln a\right)$$

As f^0 is constant at constant temperature, hence $d\ln f^0 = 0$ and the above equation gets reduced to the following:

$$\mathrm{d}G = RTd\ln a \tag{5.11}$$

Integration of this equation between the limits standard state $\left(a^0, G^0\right)$ to the existing state (a, G) yields the following:

$$\int_{G^0}^{G} \mathrm{d}G = RT\int_{a^0}^{a} d\ln a\$\,\mathrm{Or}\,\$G - G^0 = RT\left(\ln a - \ln a^0\right)$$

As pointed out above, $a^0 = 1$. Hence, the above equation becomes as follows:

$$\Delta G = G - G^0 = RT\ln a \tag{5.12}$$

This shows that from the knowledge of activity, the change in free energy in going the substance from pure state to the state of solution can be calculated by using the above equation. For i - th component, Eq. (5.12) can be written as follows:

$$\Delta G_i = G_i - G_i^0 = RT\ln a_i \tag{5.13}$$

where G_i^0 and G_i are the free energies of species i in the standard state and solution.

If the vapour above the solution is ideal, then $p_i = f_i$ and $a_i = p_i/p_i^0$. Therefore, activity of a component in a solution can be defined as the ratio of its vapour pressure exerted on the solution to its vapour pressure in pure form at the same temperature. Now, Eq. (5.13) becomes

$$\Delta G_i = RT\ln a_i = RT\ln\left(\frac{p_i}{p_i^0}\right) \tag{5.14}$$

By convention, $p_i^0 = 1$ atm. Therefore, $\Delta G_i = RT\ln p_i$ and $a_i = p_i$. The generalized formula for the relation between chemical potential and activity of a substance, showing ideal and non-ideal behaviours both, is $\mu_i = \mu_i^0 + RT\ln a_i$, where μ_i is the chemical potential of species i in the conditions of interest and μ_i^0 in the chosen standard state.

5.3 Concept of Equilibrium Constant

Let us consider the following general chemical reaction at a constant temperature and pressure:

$$aA + bB + \ldots \Leftrightarrow eE + fF + \ldots \tag{5.15}$$

where capital letters have been used to denote the reactants and products and small letters to represent their number of moles involved in the balanced form of the reaction.

The thermodynamic equilibrium constant (K) is defined as the ratio of product of activities of the chemical products to that of the reactants at equilibrium condition. In non-equilibrium condition, this ratio is known as activity quotient (Q) and is mathematically expressed as follows:

$$Q = \frac{a_E^e \cdot a_F^f \cdots}{a_A^a \cdot a_B^b \cdots}$$

When equilibrium gets established in the chemical system, the activity quotient becomes equal to equilibrium constant (i.e. $Q = K$). The logarithm of the above expression appears in the formula for ΔG^0, as discussed in the next section.

In the absence of reliable activity data, the equilibrium constant is often expressed in terms of concentrations instead of activities of reactants and products. Thermodynamically, the replacement of activity terms by concentrations is only justified for substances exhibiting ideal behaviours. Mixed units are used when the activities of components are known in one phase (e.g. the metal, etc.) but not in the second phase (e.g. the slag, etc.) of a heterogeneous reaction. In non-equilibrium state, the ratio of product of concentrations of the chemical products to that of the reactants is called concentration quotient (Q_C) which, on attainment of equilibrium, is known as equilibrium constant in terms of concentration (K_C), i.e. $Q_C = K_C$ at equilibrium. Mathematically, K_C may be expressed as follows:

$$K_C = \frac{C_E^e \cdot C_F^f \ldots}{C_A^a \cdot C_B^b \ldots} = \frac{(a_E^e/\gamma_E^e) \cdot (a_F^f/\gamma_F^f) \ldots}{(a_A^a/\gamma_A^a) \cdot (a_B^a/\gamma_B^a) \ldots} \quad (\text{Since, } a = N \cdot \gamma)$$

or

$$K_C = \frac{(a_E^e a_F^f / a_A^a a_B^a) \ldots}{(\gamma_E^e \gamma_F^f / \gamma_A^a \gamma_B^a) \ldots} = \frac{K\,(\text{equilibrium constant in terms of activity})}{\text{Quotient of activity coefficients}}$$

In the above equations, the symbols 'a' and 'γ' stand for activity and activity coefficient of respective chemical species. For ideal behaviour, the value of quotient of activity coefficients is one and, thus, $K = K_C$. Each and every system tends to acquire an equilibrium state. Here, we have the following possibilities:

Case I: If $Q_C < K_C$, the chemical reaction goes from left to right, i.e. from reactants to products. This means that movement towards the attainment of equilibrium increases Q_C by converting reactants to products.

Case II: If $Q_C > K_C$, the chemical reaction goes from right to left, i.e. from products to reactants. This means that movement towards the attainment of equilibrium decreases the value of Q_C by converting products to reactants.

Case III: If $Q_C = K_C$, the reaction mixture is at equilibrium and no chemical reaction occurs.

If we consider the chemical reaction in reverse direction, i.e. $eE + fF + \ldots \Leftrightarrow aA + bB + \ldots$, the new equilibrium constant (K'_C) becomes reciprocal of the original one (i.e. K_C) as follows:

$$K'_C = \frac{C_A^a \cdot C_B^b \cdots}{C_E^e \cdot C_F^f \cdots} = \frac{1}{K_C}$$

K_C and K'_C have different numerical values.

If the reactants and the products in Eq. (5.15) are in gaseous state, the partial pressures of the gases are used in place of activities/concentrations, and the equilibrium constant is expressed as follows:

$$K_P(\text{equilibrium constant in terms of pressure}) = \frac{p_E^e \cdot p_F^f \cdots}{p_A^a \cdot p_B^b \cdots}$$

Equilibrium constants are mainly temperature dependent, and their values remain constant at a fixed temperature for a particular reaction. Mechanism of the reaction is irrelevant in affecting the value of K. Addition of catalyst changes the mechanism and brings the system to equilibrium state more quickly, but the amounts of various chemical species are not affected. The dependence of equilibrium constant on pressure, in general, is weak in the range of pressures normally encountered in industries, and therefore, its effect is usually neglected in practice. This is true for both the condensed (i.e. solids or liquids) and gaseous phases.

There are three possible scenarios for the magnitude of equilibrium constants, and they are as follows:

Case I: If $K > 1$, the equilibrium lies to the right because concentrations of the products predominate over the reactants in the reaction system. For a very large value of K, the reaction proceeds nearly to completion.

Case II: If $K < 1$, the equilibrium lies to the left because concentrations of the reactants predominate over the products in the reaction system. For a very small value of K, the reaction proceeds hardly at all.

Case III: If $K = 1$, the equilibrium is established, and the rate of forward reaction becomes equal to that of backward reaction.

5.3.1 *Relation Between Equilibrium Constant* (K) *and Standard Free Energy Change* (ΔG^0)

The free energy change for the general chemical reaction, presented in Eq. (5.15), may be written as follows:

$$\Delta G = \sum G_{\text{products}} - \sum G_{\text{reactants}} = (eG_E + fG_F + \ldots) - (aG_A + bG_B + \ldots) \tag{5.16}$$

If the reactants and the products are in their standard states, then this equation takes the form as follows:

$$\Delta G^0 = \left(eG_E^0 + fG_F^0 + \ldots\right) - \left(aG_A^0 + bG_B^0 + \ldots\right) \tag{5.17}$$

Subtraction of Eq. (5.17) from Eq. (5.16) yields the following:

$$\begin{aligned}\Delta G - \Delta G^0 = &\left[e\left(G_E - G_E^0\right) + f\left(G_F - G_F^0\right) + \ldots\right] \\ &- \left[a\left(G_A - G_A^0\right) + b\left(G_B - G_B^0\right) + \ldots\right]\end{aligned} \tag{5.18}$$

We know that $G_i - G_i^0 = RT \ln a_i$. Use of this relationship in Eq. (5.18) leads to the following:

$$\Delta G - \Delta G^0 = RT[(e \ln a_E + f \ln a_F + \ldots) - (a \ln a_A + b \ln a_B + \ldots)]$$

$$\Delta G - \Delta G^0 = RT \ln\left(\frac{a_E^e \cdot a_F^f \ldots}{a_A^a \cdot a_B^b \ldots}\right) = RT \ln Q \tag{5.19}$$

where Q is the activity quotient. This equation has been derived by assuming the reaction to be isothermal and isobaric. When the process proceeds, the activities of reactants and the products alter, and finally, the reaction reaches equilibrium. At equilibrium, $(\Delta G)_{P,T} = 0$, and as outlined in Sect. 5.3, $Q = K$. Hence, Eq. (5.19) becomes as

$$\Delta G^0 = -RT \ln(Q)_{eq} = -RT \ln K \tag{5.20}$$

On combining Eqs. (5.19) and (5.20), we get the following:

$$\Delta G + RT \ln K = RT \ln Q \quad \text{or} \quad \Delta G = RT \ln Q - RT \ln K$$

or

$$\Delta G = RT \ln\left(\frac{Q}{K}\right) \tag{5.21}$$

Equations (5.20) and (5.21) are probably the most widely used correlations to predict the feasibility of a reaction at a particular temperature and pressure. For a chemical reaction to occur in forward direction, $(\Delta G)_{P,T} < 0$. This is satisfied only when $Q/K < 1$ or $Q < K$. ΔG is the driving force for a chemical reaction, and this has been discussed in detail in Chap. 4. Equation (5.20) can also be written as follows:

$$K = e^{-\Delta G^0/RT} = e^{\Delta S^0/R} \cdot e^{-\Delta H^0/RT} \quad (\text{Since, } \Delta G^0 = \Delta H^0 - T\Delta S^0)$$

This shows that the chemical reactions, which are accompanied by an increase of entropy (i.e. $\Delta S^0 > 0$) and decrease of enthalpy (i.e. $\Delta H^0 < 0$), have large equilibrium constant and tend to go for completion. The value of K can be calculated from the knowledge of activities of the reactants and the products. It may be worthwhile to mention here that the absolute values of the activities of substances taking part in a chemical reaction cannot be determined easily; however, it is possible to determine the value of ΔG for a reaction. Such data are available in the literature for various reactions, and from these values, one can calculate equilibrium constant at any desired temperature and pressure.

5.3.2 Temperature Dependence of Equilibrium Constant—Van't Hoff Isochore

This equation is linked with the name of J. H. Van't Hoff, who was the first to derive and use it. There are several ways to derive this equation, but the simplest method is as follows:

We know that at constant temperature and in standard state conditions of the reactants and the products, $\Delta G^0 = \Delta H^0 - T\Delta S^0$. As outlined in Eq. (5.20), $\Delta G^0 = -RT \ln K$. Hence, the equation for standard free energy change (ΔG^0) gets modified to the following:

$$-RT \ln K = \Delta H^0 - T\Delta S^0$$

$$\ln K = -\frac{\Delta H^0}{RT} + \frac{\Delta S^0}{R} \tag{5.22}$$

This is known as Van't Hoff isochore equation. Alternatively, an equation for the variation of equilibrium constant (K) with temperature can also be derived by the consideration of Gibbs–Helmholtz equation as follows:

From Gibbs–Helmholtz Eq. (4.69), we have the following:

$$\left[\frac{\partial(\Delta G^0/T)}{\partial(1/T)}\right]_P = \Delta H^0 \quad \text{or} \quad \frac{d(\Delta G^0/T)}{dT} = -\frac{\Delta H^0}{T^2} \quad \text{at constant pressure}$$

On substituting $\Delta G^0 = -RT\ln K$ from Eq. (520), the above equation becomes as

$$\frac{\mathrm{d}\ln K}{\mathrm{d}T} = \frac{\Delta H^0}{RT^2} \quad \text{or} \quad \mathrm{d}\ln K = \frac{\Delta H^0}{RT^2}\mathrm{d}T \tag{5.23}$$

On integration, we get the following:

$$\ln K = -\frac{\Delta H^0}{RT} + C \tag{5.24}$$

where C is a constant and equals to $\Delta S^0/R$. Equations (5.23) and (5.24) are the different forms of Van't Hoff equation. As already discussed in Chaps. 2 and 3, ΔH^0 and ΔS^0 are regarded as constant over small ranges of temperature because heat capacities of the products are almost equal to those of the reactants and ΔC_P is negligibly small. Hence, these equations simply show the variation of K with temperature. Over large ranges of temperatures, ΔC_P is significant, and the values of ΔH^0 and ΔS^0 of a reaction need to be evaluated by using the formula outlined in Sects. 2.4 and 3.5. Equation (5.23) can be used to calculate the value of K of any chemical reaction at any desired temperature from the knowledge of ΔH^0 of a reaction. Equation (5.24) is of the form $y = mx + c$ relationship, and plot between *ln K* (along y axis) and $1/T$ (along x axis) yields a straight line. Slope of this line is equal to $-\Delta H^0/R$ from which ΔH^0 for a reaction can be calculated, whereas an intercept of this line with $\ln K$ axis gives the value of $\Delta S^0/R$, and thus, standard entropy change (ΔS^0) for a reaction can be determined. Thus, Van't Hoff equation provides a means of calculating ΔH^0 and ΔS^0 of a reaction without performing any calorimetric or other measurements. This equation is also helpful for converting equilibrium constant data from one temperature to another, and this is done as follows:

$$\text{Attemperature}\,T_1,\ \ln K_1 = -\frac{\Delta H^0}{RT_1} + C\,\text{andattemperature}\,T_2,$$

$$\ln K_2 = -\frac{\Delta H^0}{RT_2} + C$$

As discussed above, ΔH^0 can be considered to be constant if T_1 does not differ much from T_2.

On subtracting these equations, we have the following:

$$\ln K_2 = \ln K_1 + \frac{\Delta H^0}{R}\left(\frac{1}{T_1} - \frac{1}{T_2}\right)$$

Equilibrium constant at $T_1 = 298\text{K}$ can be calculated easily from the available data of ΔG^0 at 298 K in the literature by using the relation $\Delta G^0 = -RT \ln K$. Let $T_2 = T$, then the above expression becomes as follows:

$$\ln K_T = \ln K_{298} + \frac{\Delta H^0}{R}\left(\frac{1}{298} - \frac{1}{T}\right)$$

In case of equilibria in condensed phases, equilibrium constant (K) should be expressed in terms of concentrations (i.e. K_C) and ΔH^0 needs to be replaced by ΔU^0. A very simple proof for this is given as.

We know that $\Delta H^0 = \Delta U^0 + (P_2V_2 - P_1V_1)$. Pressure and volume are both almost constant in condensed phases. Hence, $\Delta H^0 \approx \Delta U^0$. Now, the Van't Hoff isochore equation for equilibria in condensed phases can be expressed as follows:

$$\ln K_C = -\frac{\Delta U^0}{RT} + \frac{\Delta S^0}{R}$$

Equation (5.24) indicates that in order to determine the effect of temperature on the value and direction of variation of equilibrium constant, the sign and magnitude of ΔH^0 of the reaction need to be considered. Here, we have the following two possible cases:

Case I: If the reaction is endothermic, i.e. ΔH^0 is positive, the value of K increases with increase of temperature. As heat is added to such a system, the equilibrium shifts towards right, i.e. in the direction which involves the absorption of heat.

Case II: If the reaction is exothermic, i.e. ΔH^0 is negative, the value of K decreases with increase of temperature and the equilibrium shifts towards left, i.e. along reactants, leading to a decrease in the extent of forward reaction.

The net result is that the shift of equilibrium in either of the above two cases absorbs some of the added heat, and the system's temperature rise is not as large as it would be otherwise.

5.3.3 *Effect of Pressure on Equilibrium Constant in Gaseous Phase Chemical Reactions*

Majority of chemical and metallurgical processes occur at high temperatures and moderate pressures of gases (i.e. up to about 50 atmospheres). Hence, the concerned gases or gaseous mixture may be treated as ideal.

For an ideal gas mixture,

$$X_i = \frac{p_i}{P_T} \tag{5.25}$$

where X_i = mole or volume fraction of component i in the mixture and P_T = total pressure of all components, i.e. $P_T = p_1 + p_2 + \cdots + p_i + \cdots$. This equation has a straightforward application in calculation of equilibrium constant. K_P (equilibrium constant in terms of partial pressures of components) is independent of pressure because ΔG^0, being the free energy change between the products (each at unit pressure) and pure reactants (each at unit pressure), is by definition independent of pressure. But, the equilibrium constant expressed in terms of number of moles or mole fraction of the species (K_X) depends on the total pressure when there is a change in the total number of moles during a chemical reaction. The effect of change in pressure on the value of K_X can be quantified by knowing the magnitude of the change in the number of moles present in the system. For the general reaction $aA + bB = cC + dD$,

$$K_P = \frac{p_C^c \cdot p_D^d}{p_A^a \cdot p_B^b} = \frac{X_C \cdot X_D}{X_A \cdot X_B} \cdot \frac{P^c \cdot P^d}{P^a \cdot P^b} = K_X \cdot P^{(c+d)-(a+b)} \quad \text{or}$$

$$K_X = K_P / P^{(c+d)-(a+b)}$$

Now, let us examine the following possibilities:

Case I: If $(c + d) = (a + b)$, then $K_X = K_P$, i.e. the value of K_X will be independent of total pressure for no change in the number of moles during a chemical reaction.

Case II: For $(c + d) > (a + b)$, an increase of pressure exerted on the system will decrease the value of K_X and the equilibrium will shift towards left/reactants, i.e. in the direction which decreases the number of moles/effect of pressure exerted on the system.

Case III: For $(c + d) < (a + b)$, an increase of pressure exerted on the system will increase the value of K_X and the equilibrium will shift towards right/products, i.e. in the direction which decreases the number of moles/effect of pressure exerted on the system.

The concept of K_P is normally used where only gaseous components are involved, and K_X is of direct relevance in reactions involving condensed and gaseous phases. The discussions made above are in good agreement with Le Chatelier's principle which states that if pressure exerted on the reaction system (consisting of one or more gaseous components) is increased, the equilibrium will shift in that direction which tends to decrease the effect of pressure/number of moles present.

5.4 Ellingham Diagrams

Free energy change, an index to predict the feasibility of a process to occur, has already been discussed in Chap. 4. Most of the commercial processes of interest are unnatural, and temperature condition is of vital importance for their efficient operations. Hence, $\Delta G^0 - T$ diagrams/relationships are of direct use in this regard.

Basically, Ellingham diagrams, first constructed by C. J. T. Ellingham in 1944, are the graphical representations of ΔG^0 versus T relations for oxidation and sulphidation of various elements. Now, such diagrams are available not only for oxides and sulphides but also for many other compounds like chlorides, carbides, nitrides, carbonates, silicates, etc. From the oxide and sulphide Ellingham diagrams, one can have an idea about the following:

(i) Thermodynamic feasibility of oxidation and sulphidation of metals, and decomposition of metal oxides and sulphides as a function of temperature.
(ii) Thermodynamic feasibility of metallothermic and carbothermic reductions of metal oxides and sulphides as a function of temperature.
(iii) Prediction of necessary conditions for the reduction of various ores to metals.
(iv) Comparative stability of various oxides and sulphides as a function of temperature.
(v) Deoxidizers to be used in steelmaking.
(vi) Phase transformations in metal–metal oxide and metal–metal sulphide systems.
(vii) Corrosion prevention in high-temperature processes, such as nuclear power plants, where data for ΔG^0 must be available over wide ranges of temperature.
(viii) p_{H_2}/p_{H_2O} and p_{CO}/p_{CO_2} ratios as a function of temperature that can reduce a metal oxide to metal or prevent a metal from oxidation.
(ix) p_{H_2}/p_{H_2S} ratio as a function of temperature that can reduce a metal sulphide to metal or prevent a metal from sulphidation.
(x) Equilibrium partial pressures of O_2 (i.e.p_{O_2}) and S_2 (i.e.p_{S_2}) gases as a function of temperature.

The analyses derived from Ellingham diagrams are completely thermodynamic in nature and ignore the kinetics aspects. Thus, the processes predicted to be favourable by the Ellingham diagrams can still be slow.

5.4.1 Construction of Ellingham Diagram for Oxides

Ellingham diagram for the formation of various metal oxides has been presented in Fig. 5.1, and it is the most commonly and widely used diagram in metallurgy and materials science. This diagram has been drawn with $\Delta G^0 = 0$ at the top. In this oxide Ellingham diagram, ΔG^0 values for various oxide formation reactions have been plotted as a function of temperature, and all of them show similar straight lines originating from different levels (equivalent to ΔH^0 value of that reaction) on the ΔG^0 axis. The diagram plotted by Ellingham was further recast by F. D. Richardson who superimposed additional axes of po_2, CO/CO_2 ratio and H_2/H_2O ratio scales to the diagram along its right hand and bottom edges, to obtain their equilibrium values for any metal–metal oxide system over the temperature range indicated in the diagram, as shown in Fig. 5.1. The po_2 scale is used to determine the equilibrium partial pressure of oxygen at a given temperature. The significance of this is that if

po_2 is higher than the equilibrium value of po_2, the metal will be oxidized, and if it is lower than the equilibrium value, the oxide will be reduced. To use this scale, the steps required to be done are: (*i*) selection of temperature of interest, (*ii*) detection of the point where the oxidation line of interest crosses that temperature and (*iii*) lining up this point of intersection with the point 'O' marked with short radiating lines in the upper left corner of the diagram. The point, where the straight line passing through these two points crosses the *po*2 scale, is noted down, and this is the equilibrium partial pressure of oxygen. As can be seen in Fig. 5.2, the ΔG^0 value varies linearly with temperature for a fixed value of *po*2, and all these lines originate from the point 'O' in the figure which corresponds to $\Delta G^0 = 0$ at $T = 0$ K.

When carbon is used as a reducing agent, a minimum value of CO/CO_2 ratio (i.e. > equilibrium value) is required to be maintained in the reactor in order to reduce a given oxide. Poor the reducibility of an oxide, more is the amount of *CO* required in the reactor. In order to determine the equilibrium CO/CO_2 ratio at a particular temperature, the same procedure is used as outlined above for determining the equilibrium value of po_2, except lining up the point of intersection with the point 'C' marked on the centre of the left side of the diagram and noting down the point where the straight line crosses the CO/CO_2 ratio scale. Simple extrapolations of low-temperature data to high temperatures are not justified. An Ellingham diagram consists of a series of straight lines, and each line in the diagram represents a change in standard free energy or oxygen potential with temperature when one mole of oxygen gas at one atmosphere pressure reacts with a pure element to form a pure oxide. Use of one mole of oxygen gas makes all the reactions easily comparable. The equation $\Delta G^0 = \Delta H^0 - T\Delta S^0$ is of the form of straight line equation $y = mx + c$, where the intercepts of the lines with ΔG^0 or $T = 0$ K axis give the value of ΔH^0 for the concerned oxidation reactions and slope of the plot is $-\Delta S^0$

The value of ΔS^0 is negative and approximately equal for oxidation of all the metals because oxidation of each metal involves its reaction with the same quantity (i.e. one mole) of oxygen gas [M (s) + O_2 (g) = MO_2 (s)] and disappearance of the gas phase in the product side. Therefore, slopes of all the metal–metal oxide lines (i.e. $-\Delta S^0$) are positive (i.e. upward slopes) and essentially remain constant with temperature until a phase change occurs. A notable exception of this is the oxidation of solid carbon. All these have been discussed in the subsequent sections. The reactions of majority of the solid metals with oxygen are exothermic; hence, their ΔH^0 values are negative, as shown in Fig. 5.1 on the ΔG^0 axis.

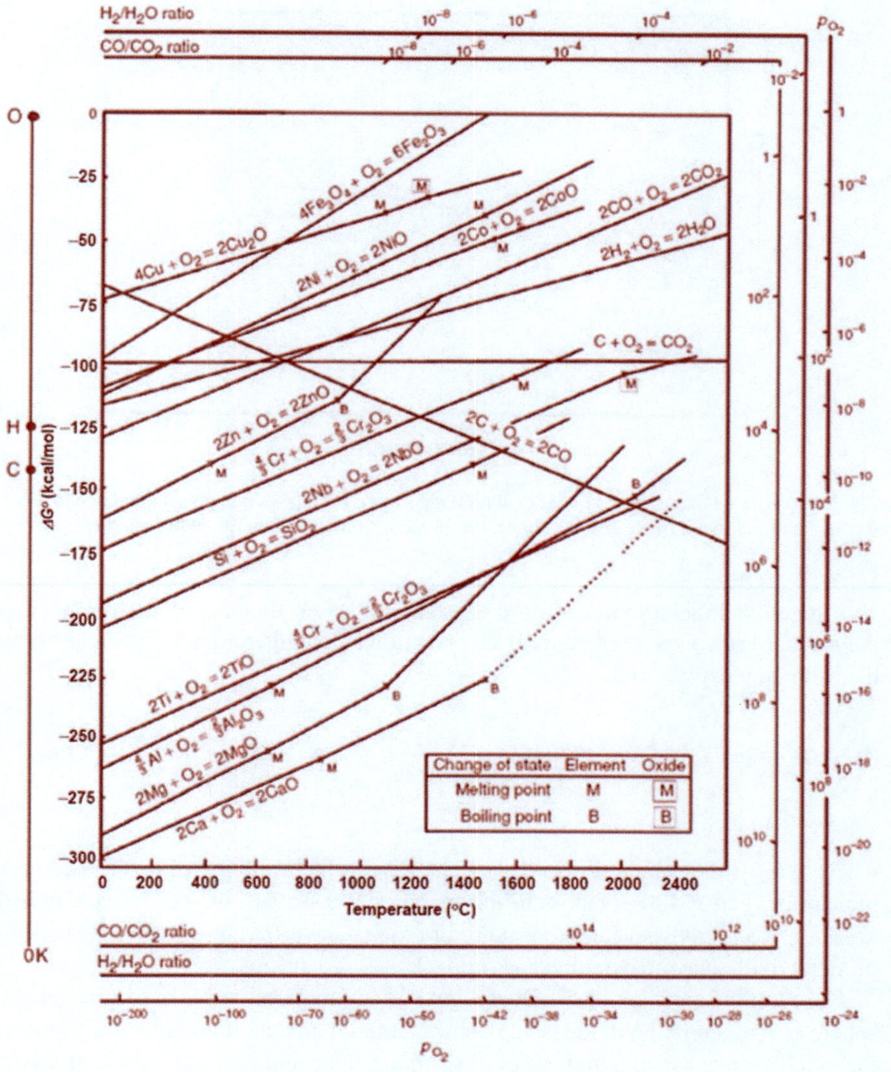

Fig. 5.1 Ellingham diagram (ΔG^0 vs. T) for oxide systems

5.4.2 Equilibrium Partial Pressure of Oxygen and Oxidation–Reduction Phenomena in a Metal–Metal Oxide System

Let us consider the following metal oxidation reaction at temperature T:

$$M(s) + O_2(g) = MO_2(s)$$

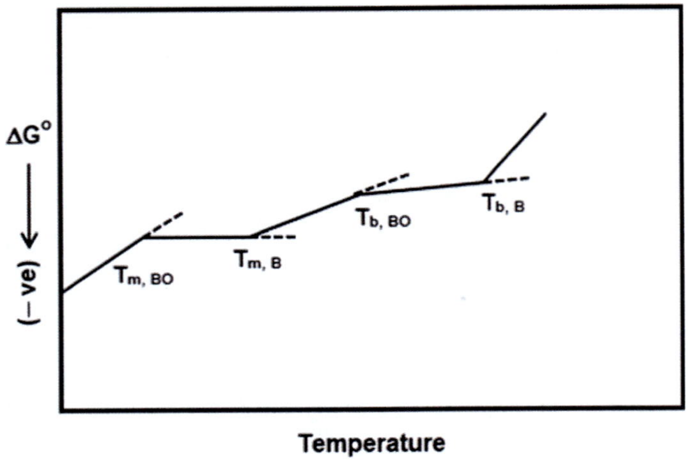

Fig. 5.2 Variation of free energy change with temperature for phase changes taking place in metal B (reactant) and metal oxide BO (product): the subscripts *m* and *b* denote melting and boiling

If metal and metal oxide are in their standard states, then $a_M = a_{MO_2} = 1$, and the standard free energy change for this reaction at a temperature T can be given by the following expression:

$$\Delta G^0 = -RT \ln K = -RT \ln\left(\frac{a_{MO_2}}{a_M \cdot p_{O_2}}\right) = -RT \ln\left(\frac{1}{p_{O_2}}\right) = RT \ln p_{O_2} \tag{5.26}$$

where ΔG^0 is the standard free energy change for the formation of metal oxide per mole of oxygen and p_{O_2} is the equilibrium partial pressure of oxygen gas and is also known as oxygen dissociation pressure of oxide. The effect of p_{O_2} on the stability of metal oxide may be examined as follows:

Case I: If the metal oxide has a very low p_{O_2} as compared to that of external atmosphere or oxygen in the oxide is at lower potential than that of an outer atmosphere, then ΔG^0 is negative and thus the oxide is more stable and cannot be reduced easily. It is also clear from Fig. 5.2 that for $p_{O_2} < 1$ atmosphere, the lines move towards more negative values of ΔG^0 (i.e. slope increases) with decrease of p_{O_2} or the oxide becomes more stable with decrease of p_{O_2} value.

Case II: If the metal oxide has higher p_{O_2} as compared to that of external atmosphere or oxygen in the oxide is at higher potential than that of an outer atmosphere, then ΔG^0 is positive and the reaction in backward direction will be feasible. In this case, the metal oxide is more unstable and can easily be reduced. Hence, in order to find out the possibility of dissociation of a metal oxide under the given set of conditions, one must know the p_{O_2} of metal–metal oxide system.

Case III: At $p_{O_2} = 1$ atmosphere, $\Delta G^0 = RT \ln p_{O_2} = 0$ at any temperature and equilibrium is established in the system. In other words, the equilibrium partial pressure of oxygen is the pressure at which the driving force for the reaction is zero. In case of equilibrium condition, neither the oxidation of metal nor the reduction or dissociation of metal oxide occurs.

From Eq. (5.22), we know that $\ln K = -\frac{\Delta H^0}{RT} + \frac{\Delta S^0}{R}$, and as mentioned above, $K = \frac{1}{p_{O_2}}$.

Hence, $\ln K = -\frac{\Delta H^0}{RT} + \frac{\Delta S^0}{R} = \ln\left(\frac{1}{p_{O_2}}\right) = -\ln p_{O_2}$ Or $p_{O_2} = e^{\frac{\Delta H^0}{RT}} \cdot e^{-\frac{\Delta S^0}{R}} = Constant \cdot e^{\frac{\Delta H^0}{RT}}$.

Here, we have the following two possibilities:

Case I: If ΔH^0 is a negative quantity (i.e. for an exothermic reaction), then p_{O_2} increases with increase of temperature. On the other hand, at constant temperature, p_{O_2} undergoes a decrease in its value as ΔH^0 becomes more negative or exothermicity of the process increases.

Case II: If ΔH^0 is a positive quantity (i.e. for an endothermic reaction), then the value of p_{O_2} decreases with increase of temperature, and at constant temperature, p_{O_2} increases with increase in endothermicity of the process, i.e. as ΔH^0 becomes more positive.

For the considered oxidation reaction M (s) + O_2 (g) = MO_2 (s), the change in entropy is

$$\Delta S^0 = S^0_{MO_2} - S^0_M - S^0_{O_2}$$

Entropy values of gases are much greater than those of condensed phases (i.e. solids or liquids), i.e. $S^0_{O_2} \gg S^0_M$ or $S^0_{MO_2}$. Hence, $\Delta S^0 \cong -S^0_{O_2}$. This indicates that standard entropy changes of all the metal oxidation reactions involving solid phases are approximately identical, and as a result, the metal–metal oxide lines in the Ellingham diagram have nearly equal and positive slopes and thus run almost parallel to each other. As mentioned in Sect. 5.4.1, this is due to the involvement of one mole of oxygen gas in all these oxidation reactions and disappearance of the gas phase in the product side.

Now, let us consider the oxidation of silver by the following reaction:

$$4Ag(s) + O_2(g) = 2Ag_2O(S)$$

For this reaction, ΔV (change in volume) is negative and $\Delta S^0 \cong -S^0_{O_2}$. Hence, the slope of the Ag – Ag_2O line, which is equal to $-\Delta S^0$, is positive (i.e. upward slope), as shown in Fig. 5.1. It can be seen in this figure that ΔG^0 value for this reaction is zero at 462 K, i.e. at this temperature, pure solid silver and oxygen gas are in equilibrium with pure solid silver oxide. From Eq. (5.26), $\Delta G^0 = RT \ln p_{O_2} = 0$ at 462 K, and therefore, p_{O_2} at this temperature is equal to 1 atmosphere. It is also evident from the diagram that as the temperature of the Ag – O_2 – Ag_2O system is decreased from 462 K to any temperature T_1 (say), ΔG^0 value for the oxidation becomes negative (i.e. Ag_2O becomes more stable than Ag and O_2 at atmosphere pressure) and the Ag spontaneously oxidizes until p_{O_2} gets reduced from

1 atmosphere to equilibrium value at which equilibrium is restored. The value of ΔG^0 at temperature T_1 is calculated by using equation $\Delta G^0_{T_1} = RT_1 \ln p_{O_2}$. Since $\Delta G^0_{T_1}$ is negative, hence, the value of p_{O_2} at temperature T_1 is less than unity. Similarly, if the temperature of the system is increased above 462 K to any temperature T_2 (say), ΔG^0 value for the oxidation reaction becomes positive (i.e. the oxide product becomes less stable than the reactants) and Ag_2O decomposes into Ag and O_2 until p_{O_2} is increased to equilibrium value at which reaction equilibrium is established.

The general concept is that as temperature rises, the ΔG^0 values become less negative and the oxide lines may intersect the horizontal $\Delta G^0 = 0$ line (i.e. $p_{O_2} = 1$ atmosphere) at some temperatures. At $\Delta G^0 = 0$, there is an equal tendency for the metal to oxidize or metal oxide to decompose, i.e. there is an establishment of equilibrium at $\Delta G^0 = 0$. If the oxidation reaction temperature is sufficient for the oxide formation line to cross the horizontal $\Delta G^0 = 0$ line, the value of ΔG^0 becomes positive and the metal oxide spontaneously decomposes to metal. The temperature at which the oxide formation line crosses the $\Delta G^0 = 0$ line is called the decomposition temperature of an oxide.

For example, the oxidation reaction lines of Hg, Pd, Fe_3O_4 and Ni intersect the $\Delta G^0 = 0$ line at the following temperatures:

$2Hg(l) + O_2(g) = 2HgO(S)$ at around 500 °C.

$2Pd(s) + O_2(g) = 2PdO(S)$ at around 900 °C.

$4Fe_3O_4(s) + O_2(g) = 6Fe_2O_3(S)$ at around 1490 °C.

$2Ni(l) + O_2(g) = 2NiO(l)$ at around 2400 °C (not shown in the diagram).

Above each of these temperatures, the overall tendency is in favour of the decomposition (i.e. reduction) of metal oxide because ΔG^0 is positive and the metal oxide is less stable than the metal. Below decomposition temperature, the feasibility of oxidation of metal increases with lowering of temperature because ΔG^0 becomes more negative. Oxidation reactions are exothermic in nature, and they always become more feasible at lower temperatures.

In fact, decomposition is the ideal process of metal extraction from its oxide merely by heating, if it is possible to separately collect the metal so formed. For example, above 462 K, silver oxide dissociates into *Ag* and O_2 and this process can be used for extraction of silver from its oxide. Unfortunately, *Ag* is not available as oxide, at least not in natural form, to make decomposition a commercial possibility. In general, decomposition temperatures are high, and it may not be possible to collect the metals in pure forms because of their highly reactive nature at high temperatures. The end result may be that the oxide dissociates into metal and oxygen at some temperature, and immediately thereafter, it gets oxidized and produces oxide again.

5.4.3 *Relative Stabilities of Metal Oxides*

The value of ΔG^0 for an oxidation reaction is a measure of the chemical affinity of the metal for oxygen. As discussed before, the more negative value of ΔG^0 at any temperature gives higher stability to the resulting oxide. The standard free energy change ΔG^0 at any temperature is the sum of ΔH^0 and $-T\Delta S^0$. As illustrated in Sect. 5.4.2, ΔS^0 values of metal oxidation reactions are virtually the same. Hence, the magnitudes of ΔG^0 and consequently the relative stabilities of the oxides are strongly dependent on the values of ΔH^0 for the respective oxidation reactions. Highly exothermic reactions (i.e. more negative values of ΔH^0) give more negative values of ΔG^0 at any temperature and higher stabilities to the oxides formed. Lower the position of a metal–metal oxide line in the Ellingham diagram, more negative is its ΔG^0 value. Hence, stability of an oxide increases with lowering of its position in the diagram, i.e. the oxides occupying lower positions in the diagram are relatively more stable than the upper oxides. For example, stability of Al_2O_3 > that of SiO_2 > that of Cr_2O_3 > that of PbO. Almost all the oxides, appearing in the diagram (i.e. Figure 5.1), have negative free energies of formation over a wide range of temperature. The exceptions are the noble metals like Pt, Au, Ag and Pd.

5.4.4 Phase Transformation

The temperature ranges in which no phase change occurs either in the reactants or products of a reaction, the ΔG^0 versus T plot for an oxidation reaction remains a straight line. The diagram (Fig. 5.1) consists of a number of straight lines that change their slopes abruptly at temperatures corresponding to a phase change (i.e. melting or vapourization) either in the metal or the oxide. Though the change in slope of the line occurs at both the melting and boiling points, as marked by symbols 'M' and 'B' in Fig. 5.1, the magnitude of alteration in the slope is more significant at the boiling points. The entropy change accompanying the transformation of solid/liquid to gas (i.e. sublimation/boiling) is greater than that for the transformation from solid to liquid, and thus the slope of the line steepens at the boiling point of metal or its oxide. For example, the zinc oxidation reactions at its melting and boiling points occur as follows:

At the melting point (693 K) of Zn,

$$2\text{Zn(l)} + \text{O}_2\text{(g)} = 2\text{ZnO(S)}$$

At the boiling point (1180 K) of Zn,

$$2\text{Zn(g)} + \text{O}_2\text{(g)} = 2\text{ZnO(S)}$$

As evident from these equations, oxidation reactions at the melting and boiling points of Zn involve the consumption of one and three moles of gas in the reactant sides. As a result, $\Delta S^0 \left(\sum S^0_{\text{products}} - \sum S^0_{\text{reactants}}\right)$ becomes relatively more negative at the boiling point, and thus, the slope steepens in upward direction.

It is known to us that the enthalpy of a high temperature phase exceeds to that of a low-temperature phase by the latent heat of a phase change, i.e. $H_l^0 = H_s^0 + L_m$ and

$H_g^0 = H_l^0 + L_V$, where L_m and L_V are the latent heats of melting and vapourization. Similarly, the entropy of a high temperature phase is greater than that of a low-temperature phase by the entropy of a phase transformation. Let us consider the oxidation reaction 2B (s) + O_2 (g) = 2BO (S) for which the standard enthalpy and entropy changes are ΔH^0 and ΔS^0. Here, the following to cases may be considered.

Case I: When a metal B melts at a temperature $T_{m,B}$, its entropy is increased and the entropy change for a reaction becomes more negative. As a result, the slope of the metal–metal oxide line gets increased. One can write the following expression for free energy change on melting of metal B:

$$\Delta G_{m,B}^0 = \Delta H_{m,B}^0 - T_{m,B}\Delta S_{m,B}^0 \quad (5.27)$$

At the melting point, both the pure solid and liquid are in equilibrium with each other. Hence, at this temperature, $G_B^0(s) = G_B^0(l)$ or $\Delta G_{m,B}^0 = 0$ and Eq. (5.27) becomes

$$\Delta H_{m,B}^0 - T_{m,B}\Delta S_{m,B}^0 = 0 \text{ or } \Delta S_{m,B}^0 = \Delta H_{m,B}^0 / T_{m,B}$$

As outlined in Chap. 3, at the melting point, $\Delta H_{m,B}^0 = L_m$ (latent heat of melting). Hence, for the post-melting oxidation reaction $B(l) + O_2(g) = 2BO(s)$, the standard enthalpy and entropy changes become equal to $\Delta H^0 - \Delta H_{m,B}^0$ and $\Delta S^0 - \Delta S_{m,B}^0$. The quantities $\Delta H_{m,B}^0$ and $\Delta S_{m,B}^0$ are always positive because melting is an endothermic process. Hence, $\Delta H^0 - \Delta H_{m,B}^0$ and $\Delta S^0 - \Delta S_{m,B}^0$ are more negative quantities than ΔH^0 and ΔS^0. Therefore, the Ellingham line for the oxidation of liquid metal B to solid BO shows greater upward slope than the corresponding line for the oxidation of solid metal B to solid BO, as shown in Fig. 5.2. Slope of the line further gets increased in upward direction at the boiling point of metal because of increase in entropy (i.e.$S_g > S_l > S_s$).

Case II: When the oxide melts, the entropy of the products is increased and the overall entropy change of the reaction becomes less negative. Hence, slope of the line gets decreased. For the reaction 2B (s) + O_2 (g) = 2BO (1), the standard enthalpy and entropy changes at the melting point of oxide become equal to $\Delta H^0 + \Delta H_{m,BO}^0$ and $\Delta S^0 + \Delta S_{m,BO}^0$. The quantities $\Delta H_{m,BO}^0$ and $\Delta S_{m,BO}^0$ are the standard enthalpy and entropy changes for the oxide phase at its melting point and are positive because melting is an endothermic process. Hence, $\Delta H^0 + \Delta H_{m,BO}^0$ and $\Delta S^0 + \Delta S_{m,BO}^0$ are less negative quantities than ΔH^0 and ΔS^0, and the line for the oxidation of $B(s)$ to $BO(l)$ shows a lesser upward slope than that for the oxidation of solid metal B to solid oxide BO, as shown in Fig. 5.2. In other words, the amount of slope of the line decreases on melting of an oxide. Slope of the line further decreases (i.e. the line bends downward) when the metal oxide boils.

5.4.5 Metallothermic Reduction

In an Ellingham diagram, highly stable oxides (i.e. those having more negative ΔG^0 values) are found at the bottom and less stable oxides (i.e. those having less negative ΔG^0 values) have occupied higher positions. Therefore, a metal occupying lower position in the diagram can always reduce the oxides of other metals lying above it, i.e. a metal is able to reduce the oxide of any other metal whose ΔG^0 versus T plot lies higher in the diagram. In other words, in metallothermic reduction process, the metal used for the reduction is so chosen that the ΔG^0 value of its oxide is more negative than that of the metal oxide to be reduced. For example, the Ellingham lines for Al–Al_2O_3 and Mg–MgO systems lie below that for Ti–TiO_2 system (Fig. 5.1). Hence, Al and Mg can be used to reduce TiO_2 by metallothermic reduction process. Positioning of the lines in the diagram also indicates that Al can be used as a reducing agent for the oxides like Cr_2O_3, Fe_2O_3, etc. (i.e. $Cr_2O_3 + 2Al = Al_2O_3 + 2Cr$). Metallothermic reduction techniques are extensively used in welding of rails (i.e. thermit welding: $Fe_2O_3 + 2Al = Al_2O_3 + 2Fe$), extraction of rare metals and production of a number of ferroalloys, such as Fe–Cr, Fe–V, Fe–Ti, Fe–Nb, Fe–Zr and Fe–Mo.

Some lines in the diagram (Fig. 5.1) intersect with others at some temperatures. As discussed in the previous section, majority of the metal–metal oxide lines run almost parallel until phase change occurs. In general, when phase change occurs in any of the metal–metal oxide system, the slope of the concerned line changes, and it intersects with other oxide lines. The intersection changes the relative positioning of these lines and thereby the relative stabilities of the concerned oxides. Let us assume that the two lines for the oxidation of metals X and Y (i.e. $2X + O_2 = 2XO$ and $Y + O_2 = YO_2$) intersect each other at temperature T_e, as shown in Fig. 5.3. It is clear from this figure that at temperature below T_e, metal oxide YO_2 (having lower position) is more stable than the oxide XO, and at temperature above T_e, the reverse is true. Hence, the reduction of metal oxide XO by metal Y is possible at temperature below T_e, while at temperature above T_e, reduction of metal oxide YO_2 by metal X is feasible.

In short, the temperature where oxide formation lines intersect, the reducing power of the elements concerned gets reversed. For example, Mg–MgO line intersects Al–Al_2O_3 line at a temperature of 1550 °C, as shown in Fig. 5.1. Hence, Mg can reduce Al_2O_3 at all the temperatures below 1550 °C because of relatively higher stability (i.e. more negative ΔG^0 value) of MgO. Above this temperature, Al_2O_3 is more stable than MgO, and hence, the metal Al is capable of reducing MgO.

5.4.6 The Oxides of Carbon and Carbothermic Reduction

When carbon reacts with oxygen, it forms two gaseous oxides, such as CO and CO_2, by the following reactions:

Fig. 5.3 Intersecting Ellingham lines for two hypothetical oxidation reactions indicating the reversal of relative stabilities after the temperature of intersection

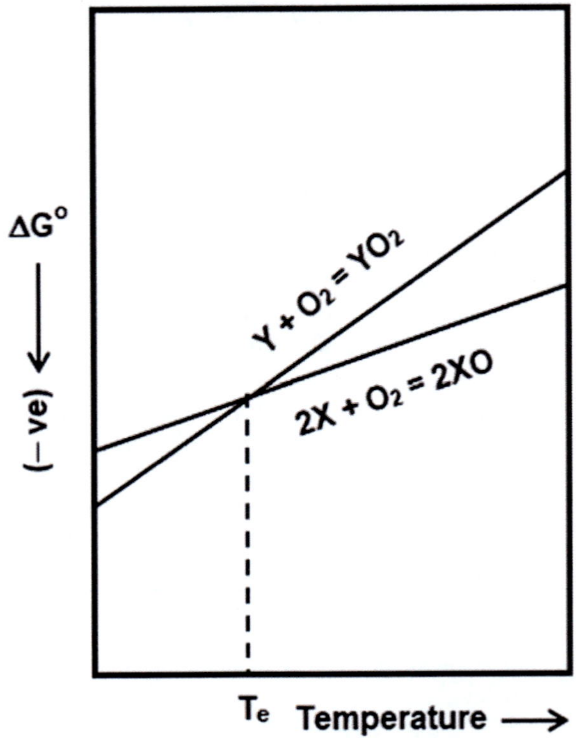

$$C(s) + O_2(g) = CO_2(g) \tag{5.28}$$

$$2C(s) + O_2(g) = 2CO(g) \tag{5.29}$$

The line for reaction (5.28) in an Ellingham diagram (Fig. 5.1) has virtually a zero slope because $\Delta V = 0$ (since one volume of CO_2 gas is produced from one volume of O_2 gas), and hence, the net change in entropy is negligibly small (i.e. $\Delta S^0 = 0$). This is the reason that C–CO_2 line is nearly horizontal and runs almost parallel to the temperature axis, i.e. the ΔG^0 value in the formation CO_2 gas is independent of temperature. The line for the reaction (5.29) has a negative or downward slope because the value of ΔS^0 is positive. This reaction is accompanied by an increase in volume (i.e.$\Delta V = +ve$) as two volume of CO gas is produced from one volume of O_2 gas. Therefore, the net change in entropy of this reaction is positive, and slope of the C–CO line in an Ellingham diagram is negative. This indicates that with rise of temperature, the ΔG^0 value in the formation of CO gas through reaction (5.29) becomes more negative, and the stability of CO gas increases in the presence of carbon.

As shown in Fig. 5.1, the C–CO_2 and C–CO lines cross each other at a temperature of around 700 °C. At this temperature, CO and CO_2 gases are in equilibrium with solid carbon. At temperatures above 700 °C, CO gas is more stable than CO_2 gas and reverse is true for temperatures less than 700 °C. The effect of intersection of lines on the stabilities of oxides has been discussed in the just previous section. The C–CO line slopes down and cuts majority of the metal oxide lines appearing in the diagram (Fig. 5.1). Therefore, carbon can reduce almost all the metal oxides, leading to the formation of CO gas, at temperatures above the points of intersection of C–CO line with these oxide lines. However, reductions of refractory oxides like TiO_2, MgO, Al_2O_3, CaO, etc. are not accomplished by using carbon because of involvement of high temperatures above 1650, 1850, 2000, 2150 °C, etc. and formation of metal carbides. Metals like Al, Ti, Cr, etc. react with carbon at the temperatures of reduction of their oxides, and the products obtained are in their carbide forms rather than pure metals. The involvement of high temperatures in the reduction of refractory oxides by carbon is also commercially not viable against other cheaper alternatives. In an upper part of the iron blast furnace (temperature range: 600–700 °C), the hematite is reduced by CO gas even in the presence of carbon. In this temperature range, 2C (s) + O_2 (g) = 2CO (g) reaction has a more negative ΔG^0 value than that for the reaction (5.29), and it has also been outlined above that CO_2 gas is more stable than CO gas at temperatures below 700 °C. Hence, CO gas acts as a stronger reducing agent than carbon.

On the other hand, H_2–H_2O line in the Ellingham diagram has a positive or upward slope which explains the decreasing stability of H_2O vapour with increase in temperature. This indicates that H_2 gas is a poor reducing agent for the reduction of metal oxides at high temperatures.

5.4.7 Deoxidation of Steel

The elements or alloys, which are added to remove the excess oxygen present in liquid steel, are called deoxidizers. Thermodynamically, following are the requirements of a good deoxidizer:

(i) It should have a higher affinity for oxygen than iron.
(ii) Its oxide should have a much lower oxygen dissociation pressure (i.e. p_{O_2}).
(iii) Its oxide must be more stable than CO gas and iron oxide at the operation temperature; i.e. ΔG^0 values of deoxidizer oxides should be more negative than those for the formation of CO gas and iron oxide.
(iv) Its oxide should not get reduced by carbon at the operation temperature.
(v) Practically, a high activity or concentration of the deoxidizer in the steel needs to be avoided because it adversely affects the properties of the deoxidized steel or its alloys.

Ellingham diagram (Fig. 5.1) gives useful information about the relative oxidizing tendencies of the deoxidizers to be used and the stabilities of the oxides formed. All elements having positions below the iron–iron oxide line in the diagram can be used as deoxidizers. On moving in downward direction from the iron–iron oxide line in the diagram, the stability of an oxide increases and deoxidation power of the concerned element, in general, increases. The deoxidation power of an element increases in the sequence Mn, Cr, V, Si, B, Ti, Al, Zr, etc., rendering the later three (i.e. Ti, Al and Zr) to be more effective in very small concentrations. In practice, Al, Mn and Si in combination are the most commonly used elements in deoxidation of steel. Thermodynamically, elements like Zr, Ti, B, V, etc. are very effective deoxidizers because of their lower positions in the Ellingham diagram, but they are more costly and rarely used. Hence, one can say that the middle 1/3rd of the Ellingham diagram constitutes the elements to be extensively used in the deoxidation of steel.

In addition to the emphasis on the stabilities of deoxidizer oxides, Ellingham diagram also gives an idea about the amounts of heat released during the deoxidation reactions of these elements. This is of primary interest in the autogenous steelmaking processes in which the necessary heat is derived from the oxidation of elements in solution. The ΔH^0 values for the deoxidizers occupying lower positions in the diagram are, in general, more negative, and hence, they release more amount of heat energy during their oxidation.

5.5 Ellingham Diagram for Sulphides

The free energy vs. temperature diagram for sulphides is important in metal extraction because many metals exit in nature in the form of sulphides. When a metallic sulphide is formed by the reaction $2M + S_2 = 2MS$, heat is released (i.e. ΔH^0 is negative) which can be calculated by the intercept of the concerned line with ΔG^0. Figure 5.4 represents the relationship between free energy change ($\Delta G^0 = RT \ln p_{S_2}$) in the formation of sulphides by consuming 1 g mole of S_2 and temperature. Comparison of ΔG^0 data cited in Figs. 5.1 and 5.4 clarifies that the values of free energy change in the formation of oxides are more negative than those for the formation of sulphides of same metals, indicating that the affinities of metals for sulphur are considerably lower than those for oxygen (by almost a factor of 1.5–2). The points S and H on the left-hand side and the scales on the right-hand side of the diagram (i.e. Fig. 5.4) can be used to calculate the values of p_{s_1} (dissociation pressure of sulphide) and H_2S/H_2 ratio of the gas mixture in equilibrium with the metal and its sulphide. The method is similar to that described for an oxide.

5.5.1 *Relative Stabilities of Sulphides*

Figure 5.4 gives a clear picture of the relative stabilities of various sulphides. Metal sulphides occupying lower positions in the diagram have more negative ΔG^0 values and are highly stable, and thus, the affinities of the concerned metals for sulphur are higher. For example, metal sulphides like CaS, CeS, BaS, MgS, Al_2S_3, etc. have occupied positions at the bottom of the diagram, and hence, they are more stable

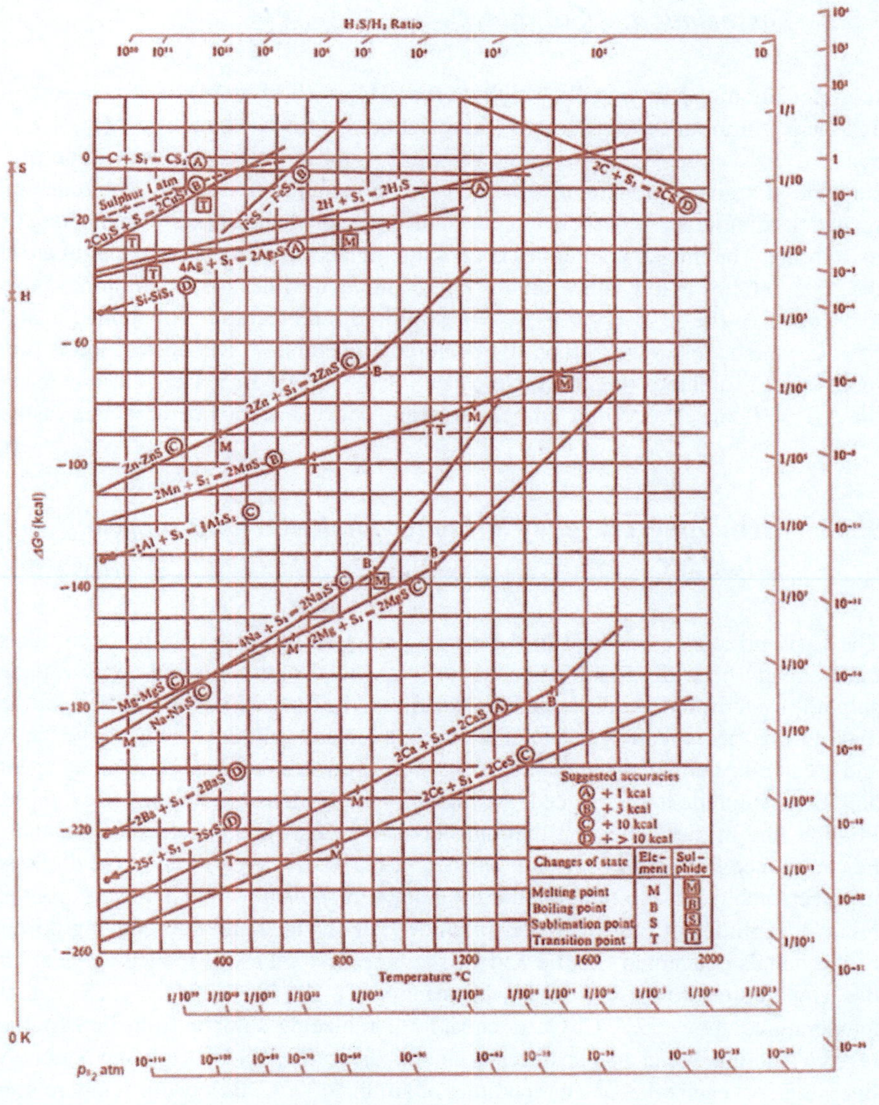

Fig. 5.4 Ellingham diagram for sulphide systems

than the sulphides appearing in the upper part of the diagram (e.g. FeS_2, CuS, Ag_2S, FeS, PbS, etc.). The higher affinity of calcium for sulphur has made it an effective disulphurizer during iron and steelmaking operations. As is evident from Fig. 5.4, the stability of the metal sulphide decreases with increase of temperature.

5.5.2 Dissociation of Sulphides

In dissociation of pure metallic sulphide (i.e. 2MS = 2 M + S_2), heat is absorbed and the equilibrium constant for this dissociation reaction is equal to p_{S_2} (i.e. $K_p = a_M^2 \cdot p_{S_2}/a_{MS}^2 = p_{S_2}$). Equilibrium pressure of sulphur vapour in dissociation of a sulphide at a given temperature is called the dissociation pressure of sulphide. As seen in the diagram, the dissociation pressure of a sulphide decreases with increase of its stability. The lines for sulphides occupying higher positions in the diagram cross the $\Delta G^0 = 0$ line at low temperatures. For example, the lines for the sulphides FeS_2 and CuS cross the $\Delta G^0 = 0$ (i.e. $p_{S_2} = 1$ atm) line at temperatures of about 700 and 550 °C, which are known as their dissociation temperatures. The lines of lower (i.e. more stable) sulphides (e.g. CaS, Cu_2S, ZnS, Na_2S, MgS, BaS, CeS, etc.) intersect the $\Delta G^0 = 0$ line at very high temperatures, and hence, they do not dissociate easily.

5.5.3 Metallothermic and Carbothermic Reductions of Metal Sulphides

The basic principles involved in these two reduction processes for oxide systems have already been discussed in detail in Sects. 5.4.5 and 5.4.6 and apply for the sulphide systems too. As discussed in Sect. 5.4.5, the metals in the lower lines of the diagram are theoretically capable of reducing the metal sulphides lying above them, and greater the vertical distance (i.e. ΔG^0 value) between the lines of reducing agent and metal sulphide to be reduced, the higher is the tendency to do so. For example, Pb–PbS line in the Ellingham diagram lies above that for Fe–FeS line, and thus, Fe can reduce PbS into Pb. As can be seen in Fig. 5.4, Cu–Cu_2S and Fe–FeS lines intersect each other at a temperature of about 625 °C. Above this temperature, the relative stabilities of the concerned sulphides get changed and the reducing power of the metals concerned (i.e. Fe and Cu) gets reversed. Fe can reduce Cu_2S at all the temperatures below 625 °C by the reaction Cu_2S + Fe = 2Cu + FeS, and at temperatures above 625 °C, Cu is capable of replacing sulphur from FeS by the reverse reaction. It can also be observed in this diagram that Mg–MgS and Na-Na_2S lines intersect each other at a temperature of around 400 °C. Below this temperature, Na_2S is more stable than MgS and thus Na acts as a reducing agent for the sulphide of magnesium. The situation is just the reverse at temperatures above 400 °C.

Unlike the metal oxides, the metal sulphides are not easily reduced by carbon. C–CS and C–CS_2 lines with slightly negative slopes are placed in the uppermost portion of the Ellingham diagram, and their positions and slopes are such that they do not intersect majority of the metal–metal sulphide lines. In addition, it can also be seen in this diagram (i.e. Figure 5.4) that the ΔG^0 and p_{S_2} values of carbon sulphides are higher in comparison with most of the metal sulphides even at high temperatures, and hence, the stabilities of CS and CS_2 are extremely poor. Therefore, carbon cannot be used as a reducing agent for the reduction of metal sulphides. Hydrogen gas can also not be used as a reducing agent for the reduction of metal sulphides due to the same reasons, as discussed above for carbon.

Discussions on phase transformations, etc. in metal sulphides are exactly the same, as done for metal oxides in Sect. 5.4.4.

5.6 Salient Features of Ellingham Diagrams—A Summary

The main characteristics of the Ellingham diagrams for oxides and sulphides are as follows:

(i) Several metal–metal oxide and metal–metal sulphide systems of interest to chemical and metallurgical engineering are shown in these diagrams with zero free energy change point at the top.
(ii) Standard free energy change for a reaction is expressed by the relation $\Delta G^0 = \Delta H^0 - T\Delta S^0$. The value of ΔH^0 (i.e. the standard enthalpy change in the formation of an oxide/sulphide) can be obtained from the intercept of the line on the ΔG^0 axis, and slope of the line yields the value of ΔS^0 (i.e. the standard entropy change in the formation of an oxide/sulphide).
(iii) Ellingham diagrams, plotted in compact forms for the formation of metal oxides and sulphides, provide huge amount of information on their relative stabilities, reduction and decomposition characteristics, possibilities of oxidation of elements, deoxidation and desulphurization of liquid steel, etc.
(iv) The top 1/3rd of the diagram comprises the lines of less reactive metals. The middle 1/3rd consists of the oxides of Si, Mn, Cr, etc. The relatively higher stabilities of these oxides make Si, Mn and Cr important deoxidizers in steelmaking. The lower 1/3rd of the diagram consists of elements forming very stable oxides, e.g. Al, Ca, Mg, etc. which are very difficult to reduce and thus are used as refractories.
(v) The lower oxides/sulphides are relatively more stable than those having positions in the upper portions of the Ellingham diagrams because on coming down in the diagrams, the value of ΔG^0 in the formation of an oxide/sulphide becomes more and more negative.
(vi) Almost all the metal oxides and sulphides have negative ΔG^0 values within the considered temperature range (i.e. 0–2200 °C) in the diagrams. The exceptions are the oxides and sulphides of noble metals like Pt, Au and to some extent Ag, which are unstable and at the top of the diagrams. This noble character decreases on coming down in the diagrams, and the most stable oxides and sulphides of base metals occupied positions at the bottom. This is nearly similar to the arrangement of metals in electrochemical series from noble to base metals in downward direction, but not exactly matching the same order.

(vii) The lines for most of the metals in the Ellingham diagrams are nearly parallel. However, some of these lines are having kinks, which correspond to phase changes in metal or its oxide/sulphide due to melting, boiling and sublimation. An alteration in the slope of the line is more significant at boiling and sublimation points than at the melting point because the values of ΔS^0 in the first two cases are higher than that at the melting point.

(viii) An element can reduce the oxide/sulphide of any other element whose ΔG^0 versus T line lies above it. The distance between the two lines considered for the reaction at a given temperature represents the standard free energy change for the reaction. The temperature where oxide/sulphide formation lines intersect, the reducing power of the elements concerned gets reversed due to change in stabilities of their oxides/sulphides.

(ix) The C–CO line in an Ellingham diagram for oxides is exceptional in that it slopes in the downward direction which is just reverse to almost all other oxides. In other words, the value of ΔG^0 for the reaction 2C (s) + O_2 (g) = 2CO (g) goes on becoming more and more negative (i.e. the stability of CO gas increases) with rise of temperature. The C–CO line cuts many of the oxide lines, and this makes carbon an useful reducing agent for the reduction of these metal oxides at temperatures above the points of intersection of this line with metal–metal oxide lines. In other words, the downward slope of C–CO line shows an increasing effectiveness of carbon as a reducing agent for the reduction of metal oxides, leading to the formation of CO gas, with rise of temperature.

(x) The line representing the reaction C (s) + O_2 (g) = CO_2 (g) is approximately parallel to the temperature axis, i.e. ΔG^0 for this reaction does not change with temperature, because this reaction is accompanied without any change in volume and entropy (i.e. $\Delta V = 0$ and A at a temperature of around 700 °C. At temperatures above 700 °C, CO is more stable than CO_2 gas, and thus reduction of any metal oxide by carbon is accompanied by the formation of CO gas, whereas at temperatures below 700 °C, reduction of any metal oxide by carbon produces CO_2 gas.

5.7 Limitations of the Ellingham Diagrams

Although, the Ellingham diagrams have been found to be very useful in estimating the feasibilities of various processes like oxidation and sulphidation of metals, deoxidation and desulphurization of liquid steel, reduction of metal oxides and sulphides, phase transformations, etc., however, they suffer from the following limitations:

(i) The free energy changes shown in the diagram are for the reactants and products when present in standard states only, i.e. in pure forms at one atmosphere pressure. But, the reactants and the products are rarely present in standard states.

(ii) The diagrams provide no information about the extent of distribution of reactants and the products between the different phases (i.e. solid, liquid or gas).
(iii) The diagrams do not provide any information about the possibility of formation of intermetallic compounds between the reactants and the products.
(iv) This diagram gives us an idea about the feasibility of oxidation, sulphidation, decomposition and reduction reactions but is totally silent about the rates of these reactions.
(v) This diagram gives a lot of knowledge about the reactants and final products obtained as a result of above-mentioned reactions but does not outline any idea about the formation of intermediate compounds.
(vi) An effect of temperature on the feasibility of these reactions has been included only in this diagram, whereas nothing was outlined about the effects of other process parameters like pressure, particle size, crystal structures, time, etc.

Solved Problems

Problem 5.1 For the decomposition reaction $CaCO_3$ (s) = CaO (s) + CO_2 (g), the value of standard free energy change is (ΔG^0) = 40,250 – 34.4 T cal/g mole. Determine the temperature at which this decomposition reaction will not occur in an open atmosphere. Also, determine the temperature if the above reaction occurs under a vacuum of 10^{-5} atm.

Solution

The given reaction is $CaCO_3$ (s) = CaO (s) + CO_2 (g)
Now, equilibrium constant$(K) = \frac{a_{CaO} \cdot p_{CO_2}}{a_{CaCO_3}}$.
As CaO and $CaCO_3$ are pure, $a_{CaO} = 1$ and $a_{CaCO_3} = 1$. Hence, $K = p_{CO_2} = 1$.
We know that $\Delta G^0 = -RT \ln K = 0$

$$\text{or } 40{,}250 - 34.4T = 0 \text{ or } T = 1170\,\text{K} = 897\,^\circ\text{C}$$

Now, under a vacuum of 10^{-5} atm, $K = p_{CO_2} = 10^{-5}$.
and $\Delta G^0 = -RT\ln K = -1.987 \times T \times \ln 10^{-5} = 22.88\ T.$
or 40,250 – 34.4 T = 22.88 T or T = 702.69 K = 429.69 °C. This indicates that a decrease in ambient pressure reduces the decomposition temperature.

Problem 5.2 The values of standard enthalpy and entropy changes at 27 °C (300 K) and 927 °C (1200 K) for the equilibrium reaction ZnO(s) + CO(g) $\rightleftharpoons$ Zn(s or g) + CO_2(g) are as follows:

$\Delta H^0_{300} = 15.55$ kcal/mol; $\Delta S^0_{300} = 3.28$ cal/Kmol and $\Delta H^0_{1200} = 43.28$ kcal/mol; $\Delta S^0_{1200} = 69.04$ cal/Kmol. By considering all the reactants and products in their standard states, determine the direction in which the above reaction is feasible at (a) 300 K and (b) 1200 K. Also, calculate the values of equilibrium constant for the above reaction at each of these temperatures.

Solution

We know that $\Delta G^0 = \Delta H^0 - T\Delta S^0$.

(a) $\Delta G^0_{300} = \Delta H^0_{300} - T\Delta S^0_{300} = 15.55 \times 10^3 - 300 \times 3.28 = 14{,}566\,\text{cal/mol} = 60{,}886\,\text{J/mol}$

(b) $\Delta G^0_{1200} = \Delta H^0_{1200} - T\Delta S^0_{1200} = 43.28 \times 10^3 - 1200 \times 69.04 = -39{,}568\,\text{cal/mol} = -165{,}394\,\text{J/mol}$

Since, the value of ΔG^0_{300} is positive, backward reaction is feasible at 300 K. However, forward reaction is feasible at a temperature of 1200 K, as the value of ΔG^0_{1200} is negative.

We know that $\Delta G^0 = -RT\ln K$.

At $T = 300$ K, $60{,}886 = -8.314 \times 300 \ln K$ or $\ln K = -24.41$ or $K = 2.5 \times 10^{-11}$.

At $T = 1200$ K, $-165{,}394 = -8.314 \times 1200 \ln K$ or $\ln K = 16.58$ or $K = 1.59 \times 10^7$.

Problem 5.3 For the chemical reaction Cu_2O (s) + H_2 (g) = 2Cu (s) + H_2O (g) occurring at a temperature of 1200 K, determine the value of equilibrium constant with the help of the following data:

$$\Delta H^0_{1200} = -22.35\,\text{kcal/mol and } \Delta S^0_{1200} = 6.96\,\text{cal/Kmol.}$$

Solution

We know that $\Delta G^0_{1200} = \Delta H^0_{1200} - T\Delta S^0_{1200} = -22.35 \times 10^3 - 1200 \times 6.96 = -30{,}702\,\text{cal/mol}$.

Hence, the reaction is feasible in forward direction.

Now, $\Delta G^0 = -RT \ln K$ or $\ln K = 30{,}702/1.987 \times 1200 = 12.88$ or $K = 3.92 \times 10^5$.

Problem 5.4 The value of ΔG^0 involved in the formation of silica (SiO_2) at 773 K, 1273 K and 1773 K is as follows:

$\Delta G^0_{773} = -175\,\text{kcal/mol}$, $\Delta G^0_{1273} = -153\,\text{kcal/mol}$, $\Delta G^0_{1773} = -131\,\text{kcal/mol}$.

Determine the values of its oxygen equilibrium partial pressures at these temperatures.

Solution

The chemical reaction is Si (s) + O_2 (g) = SiO_2 (s)

Now, $K = \frac{a_{SiO_2}}{a_{Si} \cdot p_{O_2}}$, Since $a_{SiO_2} = 1$ and $a_{Si} = 1$, hence $K = \frac{1}{p_{O_2}}$.

We know that $\Delta G^0 = -RT \ln K = -RT \ln \frac{1}{p_{O_2}} = RT \ln p_{O_2}$.

At $T = 773$ K, $\ln p_{O_2} = -175{,}000/1.987 \times 773 = -113.94$ or $p_{O_2} = 3.28 \times 10^{-50}$ atm.

At $T = 1273$ K, $\ln p_{O_2} = -153{,}000/1.987 \times 1273 = -60.49$ or $p_{O_2} = 5.36 \times 10^{-27}$ atm.

At $T = 1773$ K, $\ln p_{O_2} = -131{,}000/1.987 \times 1773 = -37.18$ or $p_{O_2} = 7.13 \times 10^{-17}$ atm.

Problem 5.5 Calculate the value of equilibrium constant (K) and equilibrium partial pressure of hydrogen for the reaction 3Fe (s) + $4H_2O$ (g) = Fe_3O_4 (s) + $4H_2$ (g) at 900 °C (1173 K). Data given are as follows:

$$\Delta G^0_{1173} = -14.76\,\text{kJ/mol and } p_{H_2O} = 0.0065\,\text{atm}$$

Solution

On using the formula, $\Delta G^0 = -RT \ln K = -R \ln \frac{a_{Fe_3O_4} \cdot p^4_{H_2}}{a^3_{Fe} p^4_{H_2O}} = -RT \ln \frac{p^4_{H_2}}{p^4_{H_2O}}$.

Because $a_{Fe} = 1$ and $a_{Fe_3O_4} = 1$.

or

$$-14{,}760 = -8.314 \times 1173 \times \ln \frac{p^4_{H_2}}{(0.0065)^4}$$

or

$$\ln p^4_{H_2} = 1.513 - 20.144 = -18.631$$

or

$$p_{H_2} = 0.0095\,\text{atm}$$

Problem 5.6 Determine the value of ΔG^0 and comment on the feasibility for the reaction PbO + Fe = Pb + FeO at 1273 K by considering the following chemical reactions and their ΔG^0 values.

$$2\text{Fe} + O_2 = 2\text{FeO},\ \Delta G^0 = -87.08\,\text{kcal/mol} \quad \text{(P.5.5.1)}$$

$$2\text{Pb} + O_2 = 2\text{PbO},\ \Delta G^0 = -44.02\,\text{kcal/mol} \quad \text{(P.5.5.2)}$$

Solution

On subtracting Eq. (P.5.2) from Eq. (P.5.1) and dividing both sides by 2, we get the following:

$$\frac{1}{2}(2\text{Fe} + O_2 - 2\text{Pb} - O_2) = \frac{1}{2}(2\text{FeO} - 2\text{PbO}),\ \Delta G^0 = \frac{1}{2}(-87.08 + 44.02)$$

$$\text{Fe} + \text{PbO} = \text{Pb} + \text{FeO}, \quad \Delta G^0 = -21.53\text{kcal/mol} = -89.99\text{kJ/mol}$$

The value of ΔG^0 is negative. Hence, the reaction is feasible in forward direction.

Problem 5.7 Determine the value of standard free energy change (ΔG^0) for the reaction NiO (s) + Co (s) = Ni (s) + CoO (s) at 327 °C. The following data are given:

ΔH^0_{298} for CoO (s) = − 57.18 kcal/mol, ΔH^0_{298} for NiO (s) = − 58.52 kcal/mol
S^0_{298} for Co (s) = 7.18 cal/Kmol, S^0_{298} for CoO (s) = 12.66 cal/Kmol, S^0_{298} for Ni (s) = 7.13 cal/Kmol, S^0_{298} for NiO (s) = 9.11 cal/Kmol
C_P for Co (s) = $5.12 + 3.42 \times 10^{-3}\, T - 0.21 \times 10^5\, T^{-2}$ cal/Kmol
C_P for CoO (s) = $11.55 + 2.04 \times 10^{-3}\, T + 0.41 \times 10^5\, T^{-2}$ cal/Kmol
C_P for Ni (s) = $7.8 - 0.47 \times 10^{-3}\, T - 1.34 \times 10^5\, T^{-2}$ cal/Kmol
C_P for NiO (s) = $- 5.0 + 37.61 \times 10^{-3}\, T + 3.9 \times 10^5\, T^{-2}$ cal/Kmol
Also determine the value of ΔG^0, if ΔC_P of the reaction is zero.

Solution

The chemical reaction is NiO(s) + Co(s) = Ni(s) + CoO(s).
We know that $\Delta G^0_{600} = \Delta H^0_{600} - 600 \times \Delta S^0_{600}$.
Now,

$$\begin{aligned}\Delta H^0_{298} &= \Delta H^0_{298}\mathrm{Ni(s)} + \Delta H^0_{298}\mathrm{CoO(s)} - \Delta H^0_{298}\mathrm{NiO(s)} - \Delta H^0_{298}\mathrm{Co(s)}\\ &= 0 - 57.18 + 58.52 - 0 = 1.34\,\mathrm{kcal/mol}\end{aligned}$$

By using the formula $\Delta H^0_{600} = \Delta H^0_{298} + \int_{298}^{600} \Delta C_P dT$.
Now,

$$\begin{aligned}\Delta C_P &= \sum C_P(\text{products}) - \sum C_P(\text{reactants})\\ &= (7.8 + 11.55 - 5.12 + 5.0) + (-0.47 + 2.04 - 3.42 - 37.61)\\ &\quad \times 10^{-3}T + (-1.34 + 0.41 + 0.213.9) \times 10^5 T^{-2}\\ &= 19.23 - 39.46 \times 10^{-3}T - 4.62 \times 10^5 T^{-2}\mathrm{cal/Kmol}\end{aligned}$$

On inserting the value of ΔC_P in the above equation, we get

$$\begin{aligned}\Delta H^0_{600} &= 1340 + \int\limits_{298}^{600} \left(19.23 - 39.46 \times 10^{-3}T - 4.62 \times 10^5 T^{-2}\right)\mathrm{d}T\\ &= 1340 + 19.23(600 - 298) - 19.73 \times 10^{-3}\left(600^2 - 298^2\right)\\ &\quad + 4.62 \times 10^5 (1/600 - 1/298)\\ &= 1340 + 5807.46 - 5350.7 - 780.34 = 1016.42\,\mathrm{cal/mol}\end{aligned}$$

Now, $\Delta S^0_{600} = \Delta S^0_{298} + \int_{298}^{600} \frac{\Delta C_P}{T}\mathrm{d}T$

$$\begin{aligned}\Delta S^0_{298} &= S^0_{298}\mathrm{Ni(s)} + S^0_{298}\mathrm{CoO(s)} - S^0_{298}\mathrm{NiO(s)} - S^0_{298}\mathrm{Co(s)}\\ &= 7.13 + 12.66 - 9.11 - 7.18 = 3.5\,\mathrm{cal/Kmol}\end{aligned}$$

$$\Delta S^0_{600} = 3.5 + \int_{298}^{600} \left(19.23 - 39.46 \times 10^{-3}T - 4.62 \times 10^5 T^{-2}\right)\frac{dT}{T}$$

Hence,

$$= 3.5 + 19.23 \times (\ln 600 - \ln 298) - 39.46 \times 10^{-3}(600 - 298)$$
$$+ 2.31 \times 10^5\left(1/600^2 - 1/298^2\right)$$
$$= 3.5 + 13.46 - 11.92 - 1.96 = 3.08\,\text{cal/Kmol}$$

Now, $\Delta G^0_{600} = \Delta H^0_{600} - 600 \times \Delta S^0_{600} = 1016.42 - 600 \times 3.08 = -831.58\,\text{cal/mol} = -3476\,\text{J/mol}$.

When $\Delta C_P = 0$, then $\Delta H^0_{600} = \Delta H^0_{298}$ and $\Delta S^0_{600} = \Delta S^0_{298}$.

Hence, $\Delta G^0_{600} = 1340 - 600 \times 3.5 = -760\,\text{cal/mol} = -3177\,\text{J/mol}$.

Problem 5.8 Find out the dissociation pressures of FeO (s) and Al_2O_3 (s) at 1873 K with the help of the following chemical reactions and associated data:

$$2Al(l) + 1.5O_2(g) = Al_2O_3(s) \quad \Delta G^0 = -718\,\text{kJ/mol}$$

$$Fe(l) + 0.5O_2(g) = FeO(l) \quad \Delta G^0 = -287\,\text{kJ/mol}$$

Solution

The chemical reaction for dissociation is $Al_2O_3(s) = 2Al(l) + 1.5O_2(g)$, $\Delta G^0 = 718\,\text{kJ/mol}$.

Now, $K = \frac{a^2_{Al} \cdot p^{1.5}_{O_2}}{a_{Al_2O_3}}$, $a_{Al_2O_3} = 1$ and $a_{Al} = 1$ on considering Al_2O_3 and Al to be in their standard states. Hence, $K = p^{1.5}_{O_2}$.

We know that $\Delta G^0 = -RT \ln K = -RT \ln p^{1.5}_{O_2}$.

or $\ln p^{1.5}_{O_2} = -718{,}000/(8.314 \times 1873) = -46.11$ or $p^{1.5}_{O_2} = 9.43 \times 10^{-21}$.

or $p_{O_2} = 3.83 \times 10^{-14}$ atm.

Similarly, the chemical reaction for dissociation is FeO (l) = Fe (l) + 0.5 O_2 (g), $\Delta G^0 = 287$ kJ/mol.

Now, $K = \frac{a_{Fe} \cdot p^{0.5}_{O_2}}{a_{FeO}}$, On considering FeO and Fe to be in their standard states, $a_{FeO} = 1$ and $a_{Fe} = 1$. Hence, $K = p^{0.5}_{O_2}$.

We know that $\Delta G^0 = -RT \ln K = -RT \ln p^{0.5}_{O_2}$.

or $\ln p^{0.5}_{O_2} = -287000/(8.314 \times 1873) = -18.43$ or $p^{0.5}_{O_2} = 9.91 \times 10^{-9}$.

or $p_{O_2} = 9.82 \times 10^{-17}$atm.

Problem 5.9 Determine the value of equilibrium constant K at a temperature 298 K for the reaction C (s) + $2H_2$ (g) = CH_4 (g) with the help of the following data:

$$\Delta H^0 = -17.91\,\text{kcal/mol and } \Delta S^0 = -19.20\,\text{cal/Kmol}$$

Solution

As per formula $\Delta G^0 = \Delta H^0 - T\Delta S^0 = -17910 + 298 \times 19.20 = -12{,}188.4\,\text{cal/mol}$.

Now, $\Delta G^0 = -RT \ln K$ or $\ln K = -\Delta G^0/RT = 12188.4/1.987 \times 298 = 20.58$.

or $K = 8.7 \times 10^8$.

Problem 5.10 Determine the value of CO/CO_2 ratio for the reduction reaction FeO (s) + CO (g) = Fe (s) + CO_2 (g) at 1173 K by using the following data:

(a) $C(s) + O_2(g) = CO_2(g) \quad \Delta G^0 = -94.29 - 0.2 \times 10^{-3} T \text{kcal/mol}$
(b) $2C(s) + O_2(g) = 2CO(g) \quad \Delta G^0 = -53.45 - 0.042T \text{kcal/mol}$
(c) $2Fe(s) + O_2(g) = 2FeO(s) \quad \Delta G^0 = -125.82 + 0.03T \text{kcal/mol}$

Solution

On simplification by 2 × Eq. (a) – Eq. (b) – Eq. (c), we get the following

$$2Fe(s) + 2CO_2(g) = 2FeO(s) + 2CO(g)$$

$$\begin{aligned}\Delta G^0 &= -188.58 - 0.0004T + 53.45 + 0.042T + 125.82 - 0.03T \\ &= -9.31 + 0.0116T \text{kcal/mol}\end{aligned}$$

Hence, $FeO(s) + CO(g) = Fe(s) + CO_2(g) \quad \Delta G^0 = 4.66 - 0.0058T \text{kcal/mol}$.

Now, equilibrium constant $(K) = \frac{p_{CO_2}}{p_{CO}}$.

We know that, $\Delta G^0 = -RT \ln K$ or $4.66 - 0.0058 \times 1173 = -1.987 \times 1173 \times \ln \frac{p_{CO_2}}{p_{CO}}$.

or $\ln \frac{p_{CO_2}}{p_{CO}} = 9.2 \times 10^{-4}$ or $\frac{p_{CO_2}}{p_{CO}} = 1.0009$.

Hence, $\frac{p_{CO}}{p_{CO_2}} = 0.999$.

Problem 5.11 Pure nickel is in contact with an atmosphere having 5% CO and 15% CO_2 at 1000 K. Can nickel be oxidized by this atmosphere? Data given are as follows:

(a) $Ni(s) + 0.5O_2(g) = NiO(s) \quad K_1 = 5.76 \times 10^7$
(b) $CO(g) + 0.5O_2(g) = CO_2(g) \quad K_2 = 1.68 \times 10^{10}$

Solution

On simplification by Eq. (b) – Eq. (a), we get.

NiO (s) + CO (g) = Ni (s) + CO_2 (g) $K = 1.68 \times 10^{10} - 5.76 \times 10^7 = 1674.24 \times 10^7$

Now, $\Delta G^0 = -RT\ln K = -8.314 \times 1000 \times \ln(1674.24 \times 10^7) = -195.72$ kJ/mol

Since, ΔG^0 value is negative, reduction of NiO by CO is feasible. On reversing the equation,

Ni (s) + CO_2 (g) = NiO (s) + CO (g) $\Delta G^0 = 195.72$ kJ/mol.

ΔG^0 value for oxidation of Ni is positive; hence, oxidation of Ni in the given atmosphere is not possible.

Problem 5.12 Calculate the value of equilibrium constant, K, for the reaction C (s) + CO_2 (g) = 2CO (g) at 847 °C and 1 atm pressure. The equilibrium mixture consists of 92.63% CO and 7.37% CO_2 by volume. Determine the value of partial pressure of CO_2 also if the value of partial pressure of CO changes to 10^{-4} atm.

Solution

Chemical reaction is C (s) + CO_2 (g) = 2CO (g)

As per question, $p_{CO} = 0.9236$ atm and $p_{CO_2} = 0.07337$ atm

Now, $K = \frac{p^2_{CO}}{a_C \cdot p_{CO_2}} = \frac{p^2_{CO}}{p_{CO_2}}$, as $a_C = 1$ for pure carbon

or $K = (0.9263)^2/0.07337 = 11.69$

When the value of p_{CO} becomes 10^{-4} atm, then $K = \left(10^{-4}\right)^2/p_{CO_2}$

or $11.69 = 10^{-8}/p_{CO_2}$ or $p_{CO_2} = 10^{-8}/11.69 = 8.55 \times 10^{-10}$atm.

Problem 5.13 For the reduction reaction FeO (s) + CO (g) = Fe (s) + CO_2 (g), calculate the percentage amount of CO gas required for the reduction of FeO at a temperature of 923 K and 1.5 atm pressure.

ΔG^0 for this reaction = − 5395.24 + 5.74 T cal/mol.

Solution

We know that $\Delta G^0 = -RT \ln K$ or $\ln K = -\Delta G^0/RT = (5395.24 - 5.74 \times 923)/1.987 \times 923$.

or $\ln K = 0.053$ or $K = 1.054$.

For the reaction FeO (s) + CO (g) = Fe (s) + CO_2 (g),$K = \frac{a_{Fe} \cdot p_{CO_2}}{a_{FeO} \cdot p_{CO}}$.

For pure Fe and FeO, $a_{Fe} = 1$ and $a_{FeO} = 1$. Hence,$K = \frac{p_{CO_2}}{p_{CO}}$.

As per question,$p_{CO} + p_{CO_2} = 1.5$

Let $p_{CO} =$ m, hence $p_{CO_2} = 1.5 -$ m.

Now, $K = \frac{p_{CO_2}}{p_{CO}} = (1.5 - \text{m})/\text{m}$ or $1.054 = 1.5/\text{m} - 1$.

or $2.054 = 1.5/\text{m}$ or m = 0.7303 or $p_{CO} = 0.73$atm

$$p_{CO_2} = 1.5 - 0.7303 = 0.77\text{atm}$$

% CO required for the reduction of FeO = (0.73/1.5) × 100 = 48.67%

Problem 5.14 Find out the value of equilibrium constant, K, for the reduction of NiO (s) by the reaction NiO (s) + H_2 (g) = Ni (s) + H_2O (g) at 1023 K. Data given are as follows:

(i) $\text{Ni(s)} + 0.5\text{O}_2\text{(g)} = \text{NiO(s)} \quad \Delta G^0 = -58506 + 23.57T\,\text{cal/mol}$

(ii) $\text{H}_2\text{(g)} + 0.5\text{O}_2\text{(g)} = \text{H}_2\text{O(g)} \quad \Delta G^0 = -58956.46 + 13.11T\,\text{cal/mol}$

Comment on the possibility of oxidation of pure Ni by H_2O during annealing at a temperature of 1023 K in an atmosphere having 5% H_2 and 95% H_2O by volume.

Solution

Subtraction of Eq. (ii) from Eq. (i) yields the following:

$$\text{Ni(s)} + \text{H}_2\text{O(g)} = \text{NiO(s)} + \text{H}_2\text{(g)} \quad \Delta G^0 = 450.46 + 10.46T$$

or $NiO(s) + H_2(g) = Ni(s) + H_2O(g) \quad \Delta G^0 = -450.46 - 10.46T$

Now, ΔG^0 for reduction of NiO (s) by H_2 (g) at $H_2(g)$ at 1023K $= -450.46 - 10.46 \times 1023$

$= -11151.04\,\text{cal/mol}$

Hence, reduction of NiO (s) by H_2 (g) is feasible at 1023 K.

For oxidation of solid Ni by H_2O gas, reaction involved is

$$Ni(s) + H_2O(g) = NiO(s) + H_2(g) \quad \Delta G^0 = 450.46 + 10.46T$$

Now, ΔG^0 value at 1023 K = 11,151.04 cal/mol.

Positive value of ΔG^0 indicates no feasibility of oxidation of solid Ni by H_2O gas at 1023 K

We know that $\Delta G^0 = -RT\ln K$ or $\ln K = -\Delta G^0/RT = -11{,}151.04/1.987 \times 1023 = -5.48$

or $K = 0.0042 = \left(\frac{p_{H_2}}{p_{H_2O}}\right)_{\text{equilibrium}}$ or $\left(\frac{p_{H_2O}}{p_{H_2}}\right)_{\text{equilibrium}} = 239.84.$

For 5% H_2 and 95% H_2O, $\left(\frac{p_{H_2O}}{p_{H_2}}\right)$ for gas phase = 95/5 = 19.

$\left(\frac{p_{H_2O}}{p_{H_2}}\right)$ for gas phase $\ll \left(\frac{p_{H_2O}}{p_{H_2}}\right)_{\text{equilibrium}}$. Hence, oxidation of pure Ni by H_2O is not feasible during annealing at 1023 K in an atmosphere having 5% H_2 and 95% H_2O.

Problem 5.15 Reduction of solid Cr_2O_3 by solid carbon occurs through the following reaction:

$$\frac{2}{3}Cr_2O_3(s) + \frac{18}{7}C(s) = \frac{4}{21}Cr_7C_3(s) + 2CO(g) \quad \Delta G^0 = 117.36 - 0.084T\,\text{kcal/mol}$$

$$\frac{2}{3}Cr_2O_3(s) + 2C(s) = \frac{4}{3}Cr(s) + 2CO(g) \quad \Delta G^0 = 125.22 - 0.084T\,\text{kcal/mol}$$

By calculation, comment on the followings:

(a) Whether the reduction of chromium oxide is easier than the carbide formation?
(b) Determine the minimum temperature above which both the reactions are feasible under standard conditions.

Solution

Consider the reduction reaction $\frac{2}{3}Cr_2O_3(s) + 2C(s) = \frac{4}{3}Cr(s) + 2CO(g)$

$$\Delta G^0 = 125.22 - 0.084T\,\text{kcal/mol}$$

We know that at equilibrium, $\Delta G^0 = 0$. Hence, $125.22 - 0.084\,T = 0$.

or $T = 1490.71$ K = 1217.71 °C.

Consider the carbide formation reaction $\frac{2}{3}Cr_2O_3(s) + \frac{18}{7}C(s) = \frac{4}{21}Cr_7C_3(s) + 2CO(g)$

$$\Delta G^0 = 117.36 - 0.084T\text{kcal/mol}$$

At equilibrium, $\Delta G^0 = 0$. Hence, $117.36 - 0.084\ T = 0$.
or $T = 1397.14$ K = 1124.14 °C

(a) The carbide formation reaction occurs at lower temperature than the reduction reaction. Hence, carbide formation reaction is easier or reduction reaction is not easier.
(b) By calculation, one can see that ΔG^0 values for carbide formation and reduction reaction become negative above 1397.14 K and 1490.71 K. Hence, the minimum temperatures for carbide formation and reduction of chromium oxide are above 1397.14 K and 1490.71 K.

Exercise

5.1 Find out the value of temperature above or below which the reduction of MnO by carbon through the chemical reaction MnO (s) + C (s) = Mn (s) + CO (g) becomes thermodynamically feasible at 1 atm pressure. Chemical reactions and associated data for them are as follows:

(a) $C(s) + O_2(g) = 2CO(g) \quad \Delta G^0 = -53451.2 - 41.94T$ cal/mol
(b) $2Mn(s) + O_2(g) = 2MnO(s) \quad \Delta G^0 = -190982.54 + 39.28T$ cal/mol

[**Ans**: 1693.3 K]

5.2 Calculate the Value of ΔG^0 in the Formation of PdO (s) by the Reaction Pd (s) + 0.5O_2 (g) = PdO (s) at a Temperature of 850 °C. Data Given: $p_{O_2} = 0.1$ atm

[**Ans**: – 20,541.18 J/mol]

5.3 Comment on the Stability of $Cr_{23}C_6$ (s) Formed Through the Reaction, $Cr_2O_3(s) + \frac{12}{23}C(s) = \frac{2}{23}Cr_{23}C_6(s) + \frac{3}{2}O_2(g)$ at 1227 °C. Data Given: $\Delta G^0 = 259.45 \times 10^3 - 62.96T\text{cal/mol}$.

[**Ans**: $Cr_{23}C_6$ is stable]

5.4 Reduction of pure FeO (s) by hydrogen gas at 1073 K occurs through the reaction FeO (s) + H_2 (g) = Fe (s) + H_2O (g). Calculate the percentage utilization of H_2 gas in this reduction reaction. Data given: $\Delta G^0 = 2001.91$ cal/mol.

[**Ans**: 71.87%]

5.5 Calculate the value of minimum temperature required for the reduction of PbO by solid carbon through the reaction 2PbO (s) + C (s) = 2Pb (s) + CO_2 (g) under standard condition. Data given are as follows:

$$2PbO(s) = 2Pb(s) + O_2(g) \quad \Delta G^0 = 105801 - 52.65T\text{cal/mol}$$

$$C(s) + O_2(g) = CO_2(g) \quad \Delta G^0 = -94290 - 0.201T\text{cal/mol}$$

[**Ans**: 217.78 K]

5.6 The Values of ΔG^0 in the Formation of Silica (SiO_2) at 773 K, 1273 K and 1773 K Are as Follows:

ΔG^0 at 773 K = − 175 kcal/mol, ΔG^0 at 1273 K = − 153 kcal/mol and ΔG^0 at 1773 K = − 131 kcal/mol. Calculate its equilibrium oxygen pressures (p_{O_2}) at these temperatures.

[**Ans**: 3.3×10^{-50} atm at 773 K; 5.38×10^{-27} atm at 1273 K; 7.09×10^{-17} atm at 1773 K]

5.7 Comment on the possibility of reduction of Al_2O_3 at 1773 K in an atmosphere of hydrogen gas. If not then what is the minimum temperature required? The chemical reactions and the values of ΔG^0 are given as follows:

(i). $2Al(l) + 1.5O_2(g) = Al_2O_3(s), \quad \Delta G^0 = -401884.21 + 76.98T\,\text{cal/mol}$

(ii). $H_2(g) + 0.5O_2(g) = H_2O(g), \quad \Delta G^0 = -60257.61 + 13.95T\text{cal/mol}$

[**Ans**: No possibility of reduction at 1773 K; minimum temperature required = 6295.31 K]

5.8 Calculate the minimum temperature required for the reduction of solid ZnO by carbon through the reaction ZnO (s) + C (s) = Zn (g) + CO (g). Also, find out the gas composition at the total pressures of: (i) 1 atm and (ii) 0.1 atm. Following data are given:

$$ZnO(s) = Zn(g) + 0.5O_2(g), \quad \Delta G^0 = 115.11 - 0.05T\text{kcal/mol}$$

$$C(s) + 0.5O_2(g) = CO(g), \quad \Delta G^0 = -26.42 - 0.021T\text{kcal/mol}$$

[**Ans**: 1201.36 K at 1 atm, 1068.35 K at 0.1 atm; in both cases gas composition = 50:50]

5.9 The Value of ΔG^0 for the Decomposition Reaction $ZrO_2 = Zr + O_2$ is as Follows:

$$\Delta G^0 = 260188.76 + 4.33T \log T - 59.18T\text{cal/mol}$$

Find out the following:

(a) The values of equilibrium constant and equilibrium partial pressure of oxygen at a temperature of 2000 K.
(b) Comment on the possibility of decomposition of pure zirconia (ZrO_2) crucible under a vacuum of 10^{-5} mm of Hg at the above-mentioned temperature.

[**Ans**: (a) $K = 2.459 \times 10^{-19}$; $p_{O_2} = 1.87 \times 10^{-16}$ mm of Hg, (b) no possibility of decomposition]

5.10 The Value of ΔG^0 for the Reaction C (s) + CO_2 (g) = 2CO (g) is ΔG^0 = 40,839 − 41.74 T cal/mol. Calculate the Following:

(a) The value of temperature at which graphite (C) will be in equilibrium with a gaseous mixture having (i) 30% CO and 70% CO_2 and (ii) 60% CO and 40% CO_2 (at 1 atm pressure).
(b) Similar values under the same conditions at a total pressure of 0.1 atm.

[**Ans**: (a) (i) 891.31 K; (ii) 973.55 K, (b) (i) 810.31 K; (ii) 929.95 K]

5.11 The values of free energy change (ΔG^0) involved in the formation of NiO at 727 °C and 927 °C are – 35.94 kcal/mol and – 31.92 kcal/mol. With the help of these data, determine the values of the following:

(a) Equilibrium constant at each temperature.
(b) ΔH^0 and ΔS^0.

[**Ans**: (a) $7.1 \times 10^7, 6.4 \times 10^5$; (b) $\Delta H^0 = -234.4$ kJ/mol, $\Delta S^0 = -84.1$ J/K mol]

5.12 Reduction of tungsten oxide (WO_2) by hydrogen gas to form metallic tungsten (W) at 1673 K occurs through the reaction WO_2 (s) + $2H_2$ (g) = W (s) + $2H_2O$ (g). Find out the maximum concentration of water vapour which could be permitted to be in the reactor for the above reduction reaction to occur. Data given are as follows:

$$\mathrm{W(s)} + \mathrm{O_2(g)} = \mathrm{WO_2(s)} \quad \Delta G^0 = -134929 + 38.96T\,\mathrm{cal/mol}$$

$$\mathrm{H_2(g)} + 0.5\mathrm{O_2(g)} = \mathrm{H_2O(g)} \quad \Delta G^0 = -60257.61 + 13.95T\,\mathrm{cal/mol}$$

[**Ans**: 64.84% (max)]

Chapter 6
Thermodynamics of Solutions

Solutions have thermodynamic properties distinct from those of their constituents. Solution formation is governed by the changes in volume, enthalpy, free energy, entropy, etc., and a proper knowledge of these is essential to understand the behaviours of solutions formed. The basic thermodynamic properties of ideal, real and regular solutions along with changes in enthalpy (ΔH), Gibbs free energy (ΔG) and entropy (ΔS) in their formation have been dealt with in Sects. 6.2, 6.7 and 6.9 of this chapter. Many equations in this context have been derived and discussed. A thorough discussion on Raoult's law, interaction between components in solution, positive and negative deviations from ideality, partial molar quantities and Gibbs–Duhem equation has been provided in Sects. 6.2–6.6. Gibbs–Duhem equation is very useful in evaluating activity of a component in the solution.

In addition to these, concepts of an excess function and Henry's law along with their importance have also been outlined substantially in this chapter. A short discussion has been made on 1 wt.% standard state and dilute multicomponent solutions in this chapter. Numerous problems have been solved and given as an exercise at the end of this chapter.

6.1 Introduction

In reality, the concept of existence of a perfectly pure material is hypothetical because the absolutely pure state is thermodynamically unattainable. The work impure implies that the specific species occurs in solution, either solid or liquid, with some other species. It has been well established that nearly all substances of metallurgical interest are not pure materials but solutions of various elements or compounds (e.g. pig iron, steel, matte, slag, etc.).

A solution is a single homogeneous phase consisting of different chemical substances, whose concentration may be varied without precipitation of a new phase.

S. K. Sahoo et al., *Fundamentals of Metallurgical Thermodynamics*,
https://doi.org/10.1007/978-981-99-6671-4_6

It differs from a mixture by its homogeneity and from a compound by being able to possess variable composition. The component having largest concentration in solution is referred to as solvent, and others are solutes. Majority of the metallurgical processes are usually carried out at high temperatures. Hence, all the metallurgical solutions are non-aqueous and inorganic, and they may be classified into metallic solutions (e.g. pig iron, steel, Ag-Au and Fe–Ni alloys, etc.) and non-metallic solutions (e.g. a molten slag containing CaO, FeO, SiO_2, Al_2O_3, etc.). The more important thermodynamic properties of these solutions have been discussed in this chapter. The thermodynamic properties of a solution are quite different from those of its components because of interaction between the component atoms/molecules, and this can be understood by the following examples:

Let components A and B form a solution AB. V_A and V_B—Volumes of components A and B before mixing. V_{AB}—Volume of a solution AB. It has been seen that $V_{AB} \neq V_A + V_B$, i.e. $V_{AB} > V_A + V_B$ or $V_{AB} < V_A + V_B$.

If $F_{AB} > F_{AA}$ or F_{BB}, then $V_{AB} < V_A + V_B$, i.e. contraction in volume occurs during the formation of a solution. If $F_{AB} < F_{AA}$ or F_{BB}, then $V_{AB} > V_A + V_B$, i.e. an expansion in volume occurs as a result of formation of a solution.

Where F_{AA} and F_{BB} are the attractive forces between similar atoms of components A and B and F_{AB} is the attractive force between dissimilar atoms of components A and B.

The composition of a solution is usually expressed in terms of either wt.%, atom % or mole %. The atom/mole fraction, most widely used in thermodynamic equations, is defined as the ratio of number of atoms/moles of a substance to the total number of atoms/moles of all the substances present in the solution.

6.2 Ideal and Non-ideal Solutions

An ideal solution has been defined as one whose components obey Raoult's law at all temperatures and pressures at which the solution is capable to exist. In early days, the studies on solutions were limited to the measurements of vapour pressures of components in binary solution systems. The observations so made were codified by Raoult in the form of a statement whose scope was limited to binary solutions only. Raoult's law states that the relative lowering of vapour pressure of a solvent due to the addition of a solute is equal to the mole fraction of solute in the solution. Let N_A and N_B are the mole fractions of components A and B in a solution formed by their mixing, and p_A and p_B are their vapour pressures in the existing state. p_A^0 and p_B^0 are the vapour pressures of components A and B in their pure forms at the same temperature at which the solution exists. Now, according to Raoult's law, we have the following equations:

$\frac{p_A^0 - p_A}{p_A^0} = N_B$ (if A is the solvent) and $\frac{p_B^0 - p_B}{p_B^0} = N_A$ (if B is the solvent).

or $1 - \frac{p_A}{p_A^0} = N_B$ and $1 - \frac{p_B}{p_B^0} = N_A$

or $\frac{p_A}{p_A^0} = 1 - N_B = N_A$ and $\frac{p_B}{p_B^0} = 1 - N_A = N_B$

In a generalized form, $$\frac{p_i}{p_i^0} = N_i \text{ or } p_i = N_i p_i^0 \tag{6.1}$$

Equation (i) is the expression of Raoult's law in terms of partial pressures of a component. As outlined in Sect. 5.1, $f = p$. Hence, the above equation can be written as:

$$N_i = \frac{p_i}{p_i^0} = \frac{f_i}{f_i^0} = a_i \text{ or } a_i = N_i \tag{6.2}$$

This is the another mathematical equation for the Raoult's law which states that the activity of a solute '*i*' in the solution at all the concentrations is equal to its mole fraction, and the solution is said to be ideal. Various mathematical expressions for the Raoult's law can now be summarized as follows:

(i) $p_i = N_i p_i^0$ (in terms of partial pressure of component '*I*' in the solution).
(ii) $f_i = N_i f_i^0$ (in terms of fugacity of component '*i*' in the solution).
(iii) $a_i = N_i$ (in terms of activity of component '*i*' in the solution).

No any solution completely satisfies the requirements of an ideal solution over the entire concentration range. In reality, almost all the solutions exhibit a positive or negative deviation from Raoult's law line/ideal behaviour because of attractive or repulsive interactions between component atoms. In some cases, the deviations from ideality are extremely small and may be ignored. Such solutions are treated as ideal for most practical purposes. Fe–Mn, Fe–Ni, Fe–Co, Cu–Ag, FeO–MnO and FeS–FeO solutions have been observed to exhibit near ideal behaviour. In majority of the cases, the departure from ideality/Raoult's law line is significant, and the application of ideal solution equations can give rise to serious problems.

The solutions which do not obey Raoult's law are known as non-ideal or real solutions, and their activity vs. mole fraction curves are either above or below the diagonal, i.e. Raoult's law line, as shown in Fig. 6.1. Raoult's law has been reported to hold true in dilute solutions only. In non-dilute solutions, the behaviour deviates considerably from the ideality in both the positive and negative directions. In the deviated form, the activities of the components are either more (+ve deviation) or less (−ve deviation) than the ideal values. All these have been illustrated in Fig. 6.1. The negative deviation has been reported to be exhibited by solutions like Fe–Si, Fe–P, Fe–S, Fe–Sn, Al–Zn, Fe–Al, Fe–Ti, $CaO–SiO_2$, $FeO–SiO_2$, $MgO–SiO_2$, $MnO–SiO_2$, etc. The positive deviation from ideality is shown by relatively few systems only, such as Fe–Cu and Fe–Pb. Departures from Raoult's law line/ideality have been quantified thermodynamically by using a parameter known as activity coefficient (γ) which is mathematically expressed by the following equation:

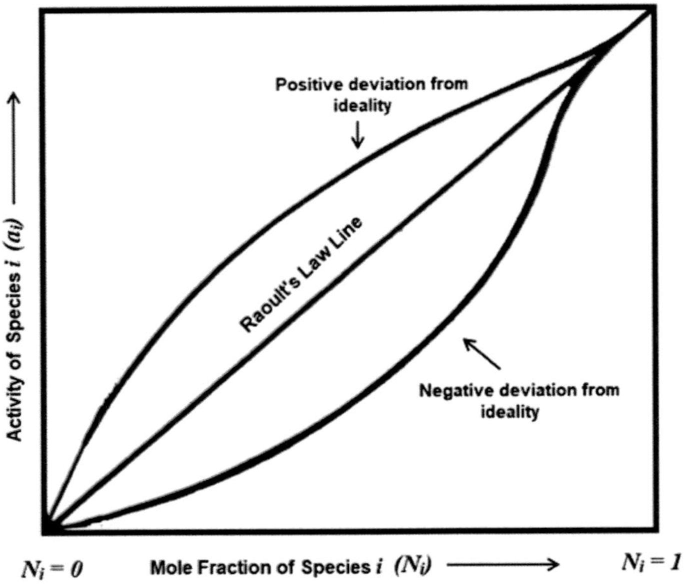

Fig. 6.1 Raoult's law line with positive and negative deviations from ideality

$$\gamma_i = \frac{a_i}{N_i} \text{ or } a_i = \gamma_i N_i \tag{6.3}$$

where γ_I, a_i and N_i are the activity coefficient, activity and mole fraction of component 'i' in the solution. Here, we have the following three cases:

Case I: When $a_i = N_i$, then $\gamma_i = 1$, and the solution is said to be ideal.
Case II: When $a_i > N_i$, then $\gamma_i > 1$, and the solution exhibits positive departure from ideality.
Case III: When $a_i < N_i$, then $\gamma_i < 1$, and the solution exhibits negative departure from ideality.

A qualitative interpretation for the cause of deviation from ideality can be obtained from consideration of the forces holding atoms or molecules in the solution, as discussed in the next coming section.

6.3 Component Interaction in Solution and Deviation from Ideality

Interaction effect on deviation from ideality can be discussed more easily if one considers a binary solution of components A and B. There is a natural tendency for the component atoms to undergo interaction with each other and to segregate into

A–A, B–B and A–B pairs. The possible interactions in this binary solution are as follows:

(i) A–B type of interaction, i.e. grouping of dissimilar or unlike atoms/molecules in the system.
(ii) A–A type of interaction, i.e. grouping of similar or like atoms/molecules of component A in the solution.
(iii) B–B type of interaction, i.e. grouping of similar or like atoms/molecules of component B in the solution.

The interactions lead to changes in the energies of the constituent particles in the solution and heat is generally liberated or absorbed depending upon the nature of interactions. ΔH (heat of mixing) value associated with A–A and B–B type of groupings is positive (i.e. endothermic), whereas A–B type of grouping generally gives rise to a negative heat of mixing (i.e. exothermic). Heat of mixing is zero for an ideal solution. Almost entire volume of solution is available for component atoms/molecules to move. Hence, randomness in the system increases, and the entropy of the components on solution formation increases. The type of interaction between component atoms determines their activities. The deviation of activities from ideal mixing can also be predicted qualitatively by considering the forces of attraction between two like atoms or unlike atoms in a solution. Now, let us consider the physical significance of deviation from ideality in terms of interactions and forces of attraction between like atoms or unlike atoms in a binary solution consisting of components A and B.

Ideal Behaviour: When the net attractive force between components A and B is equal to the average of the attractive forces between A–A and B–B, i.e. $F_{AB} = \frac{1}{2}(F_{AA} + F_{BB})$, the activities of the components A and B in the solution at all concentrations are equal to their mole fractions (i.e. $a_A = N_A$, $\gamma_A = 1$ and $a_B = N_B$, $\gamma_B = 1$), and the solution is said to be ideal. The ideal/Raoultian behaviour is exhibited by the components if their atoms are similar in nature. But, such a situation is non-existent. However, if they are almost alike, the behaviour is considered to be practically ideal. Examples of such cases are obviously few.

Positive and Negative Deviations from Ideality: When the attraction between similar atoms/molecules dominates over attraction between dissimilar atoms/molecules (i.e. A–A and B–B bonds are stronger than A–B bonds), then the components A and B exhibit positive deviation from ideality. In this case, both the components tend to form their clusters in the solution, and their activity values get increased (i.e. $\gamma_A > 1$ and $\gamma_B > 1$) and may exhibit miscibility gap, as is seen in silicate melts. In terms of forces of attraction, the positive departure from ideality may be represented as follows:

$$F_{AB} < \frac{1}{2}(F_{AA} + F_{BB})$$

The solutions showing positive departure from ideality are, in general, endothermic in nature. On the contrary, if A–B interactions are significantly higher than the average of A–A and B–B interactions $\left(\text{i.e. } F_{AB} > \frac{F_{AA}+F_{BB}}{2}\right)$, then both the components exhibit negative departure from ideality, and in consequence, intermetallic compounds like A_xB_y tend to form. This type of interaction reduces the activity values of the components (i.e. $\gamma_A < 1$ and $\gamma_B < 1$) and their escaping tendencies from the solution to the surroundings. The solutions exhibiting negative departure from ideality are usually exothermic in nature.

6.4 Partial Molar Quantities and Gibbs–Duhem Equation

Metallurgical processes are usually concerned with metals and alloys containing varying amounts of elements, such as Fe, Cu, Ni, Cr, Zn, V, C, Mn, Si, P and S and a number of compounds. The thermodynamic properties of these components in the solution differ from those in their standard states. This is due to the difference between the interatomic forces in the pure substance and in the solution. In order to assess the contribution of these components individually on the change in their thermodynamic properties (e.g. *V*, *H*, *G*, *S*, etc.) after mixing, it is necessary to introduce the concept of partial molar quantities. All the thermodynamic quantities (e.g. *V*, *H*, *G*, *S*, etc.) undergo similar change when a substance changes from pure state to a state of solution. The value of any extensive state property of a component/species in the mixture/solution per mole of that component/species is known as the partial molar value of the property. The partial molar value of an extensive thermodynamic property, *Q*, of the component '*i*' in a solution having components like *i*, *j*, *k*, …. is formally defined by the following equation:

$$\overline{Q_i} = \left(\frac{\delta Q'}{\delta n_i}\right)_{T,P,n_j,n_k,\ldots} \tag{6.4}$$

where Q_i is the value of an extensive thermodynamic property for the arbitrary quantity of the solution and $\overline{Q_i}$ represents the change in value of Q' of the solution when one mole of component '*i*' is added to the large quantity of the solution at constant *T*, *P* and mole numbers of all other components. The composition of the solution must remain virtually constant during the addition of a particular component. Hence, the quantity of the solution must be sufficiently large so that the addition of a further mole of any component to the solution makes virtually no change in the overall composition of the solution or does not change the composition of the solution appreciably or should cause no measurable change in the composition of the solution. Similarly, the partial molar values for other components *j*, *k*, …. can be written as follows:

$$\overline{Q_j} = \left(\frac{\delta Q'}{\delta n_j}\right)_{T,P,n_i,n_k,\ldots} \quad \text{and} \quad \overline{Q_k} = \left(\frac{\delta Q'}{\delta n_k}\right)_{T,P,n_i,n_j,\ldots}$$

In case of Gibbs free energy being the thermodynamic property, Eq. (6.4) for partial molar Gibbs free energy of component '*i*' in the solution can be written as follows:

$$\overline{G_i} = \left(\frac{\delta G}{\delta n_i}\right)_{T,P,n_j,n_k,\ldots} \tag{6.5}$$

We also know that $\mu_i = \overline{G_i}$ or $\overline{F_i}$ or $\overline{U_i}$ or $\overline{H_i}$, where μ_i is the chemical potential of species '*i*' in the solution and the symbols $\overline{G_i}$, $\overline{F_i}$, $\overline{U_i}$ and $\overline{H_i}$ stand for partial molar Gibbs free energy, Helmholtz free energy, internal energy and enthalpy of component '*i*' in the solution. Each of these thermodynamic properties acts as a chemical potential under different conditions to assess chemical equilibrium. G acts as a chemical potential under constant temperature and pressure conditions, F under constant temperature and volume conditions, U under constant entropy and volume conditions and H under constant entropy and pressure conditions. Most of the chemical and metallurgical processes are carried out under constant temperature and pressure conditions. Hence, partial molar Gibbs free energy $\overline{G_i}$ is commonly used as a chemical potential to assess chemical equilibrium in the solution, and thus, Eq. (6.5) can be rearranged as

$$\overline{G_i} = \left(\frac{\delta G}{\delta n_i}\right)_{T,P,n_j,n_k,\ldots} = \mu_i$$

Now, the chemical potential of species '*i*' in a solution can be defined as an increase in Gibbs free energy of the system per mole of species '*i*' added at constant *T*, *P* and number of mole of other species present in the solution.

We know that the thermodynamic property (Q') of a solution is a function of temperature, pressure and composition, i.e.

$$Q' = Q'(T, P, n_i, n_j, n_k, \ldots) \tag{6.6}$$

At constant temperature and pressure, this equation can be differentiated partially w.r.t number of moles ($n_i, n_j, n_{k,\ldots}$) of each of the component as follows:

$$\mathrm{d}Q' = \left(\frac{\delta Q'}{\delta n_i}\right)_{T,P,n_j,n_i,\ldots} \mathrm{d}n_i + \left(\frac{\delta Q'}{\delta n_j}\right)_{T,P,n_i,n_k,\ldots} \mathrm{d}n_j + \left(\frac{\delta Q'}{\delta n_k}\right)_{T,P,n_i,n_j,\ldots} \mathrm{d}n_k \tag{6.7}$$

or

$$\mathrm{d}Q' = \overline{Q_i}\mathrm{d}n_i + \overline{Q_j}\mathrm{d}n_j + \overline{Q_k}\mathrm{d}n_k \tag{6.8}$$

where $\overline{Q_i}, \overline{Q_j}, \overline{Q_k}, \ldots$ are called partial molar quantities of components *i*, *j*, *k*, … $\overline{Q_i}, \overline{Q_j}, \overline{Q_k}, \ldots$ are functions of temperature, pressure and composition of the solution, and their values will remain constant because temperature, pressure and composition are constant/fixed. Therefore, Eq. (6.8) becomes

$$Q' = \overline{Q_i} n_i + \overline{Q_j} n_j + \overline{Q_k} n_k + \ldots \tag{6.9}$$

On differentiation of this equation, we get the following:

$$\begin{aligned} \mathrm{d}Q' &= \left(\overline{Q_i}\mathrm{d}n_i + n_i\mathrm{d}\overline{Q_i}\right) + \left(\overline{Q_j}\mathrm{d}n_j + n_j\mathrm{d}\overline{Q_j}\right) + \left(\overline{Q_k}\mathrm{d}n_k + n_k\mathrm{d}\overline{Q_k}\right) + \ldots \\ \text{or}\quad \mathrm{d}Q' &= \left(n_i\mathrm{d}\overline{Q_i} + n_j\mathrm{d}\overline{Q_j} + n_k\mathrm{d}\overline{Q_k} + \ldots\right) + \left(\overline{Q_i}\mathrm{d}n_i + \overline{Q_j}\mathrm{d}n_j + \overline{Q_k}\mathrm{d}n_k + \ldots\right) \end{aligned} \tag{6.10}$$

On equating Eqs. (6.8) and (6.10), we get

$$n_i\mathrm{d}\overline{Q_i} + n_j\mathrm{d}\overline{Q_j} + n_k\mathrm{d}\overline{Q_k} + \ldots = 0$$

In general, this can be written as

$$\sum_i n_i\mathrm{d}\overline{Q_i} = 0 \tag{6.11}$$

Division of Eq. (6.11) by total number of moles of the solution gives,

$$\sum_i N_i\mathrm{d}\overline{Q_i} = 0 \tag{6.12}$$

Equations (6.11) and (6.12) are equivalent expressions of the Gibbs–Duhem equation. Equation (6.12) is a very important form of Gibbs–Duhem equations for solutions, and a number of subsequent relations have been derived from this.

Let us suppose that Q is the Gibbs free energy of the solution, then Eq. (6.12) can be written as

$$N_i\mathrm{d}\overline{G_i} + N_j\mathrm{d}\overline{G_j} + \ldots = 0 \tag{6.13}$$

where N_i, N_j, … are mole fractions of components *i*, *j*, … and $\overline{G_i}$, $\overline{G_j}$, … are their molar Gibbs free energies. Similarly, we may also write the following:

$$N_i\mathrm{d}\overline{H_i} + N_j\mathrm{d}\overline{H_j} + \ldots = 0 \text{ (for enthalpy)}$$

$$N_i\mathrm{d}\overline{S_i} + N_j\mathrm{d}\overline{S_j} + \ldots = 0 \text{ (for entropy)}$$

where $\overline{H_i}$, $\overline{H_j}$, … are partial molar enthalpies of components *i*, *j*, … and $\overline{S_i}$, $\overline{S_j}$, … are their partial molar entropies.

We also know that $dG = RTd\ln a$. On using this equation in Eq. (6.13), we get the following in terms of activities of components:

$$N_i(RT\,\mathrm{d}\ln a_i) + N_j\big(RT\,\mathrm{d}\ln a_j\big) + \ldots = 0$$

$$\text{or}\quad N_i\mathrm{d}\ln a_i + N_j\mathrm{d}\ln a_j + \ldots = 0 \tag{6.14}$$

This is another form of Gibbs–Duhem equation in terms of activities of different components in the solution. Integration of Gibbs–Duhem equation based on Darken's α-function is generally employed and is given by the expression:

$$\alpha_i = \frac{\ln\gamma_i}{(1-N_i)^2}\ \text{for } i\text{-th species in the solution} \tag{6.15}$$

For a binary solution having components A and B, $\alpha_A = \frac{\ln\gamma_A}{(1-N_A)^2} = \frac{\ln\gamma_A}{(N_B)^2}$ and $\alpha_B = \frac{\ln\gamma_B}{(N_A)^2}$.

Use of this α-function in Gibbs–Duhem equation, and integration between the limits $N_A = 1$ and $N_A = N_A$ yields the following equation for activity coefficient of component A:

$$\ln\gamma_A = -N_A N_B\alpha_B - \int\limits_{N_A=1}^{N_A} \alpha_B \mathrm{d}N_A \tag{6.16}$$

Equation (6.15) and (6.16) are frequently used in determining the values of activity coefficients of components and solving the problems.

6.5 Integral and Partial Molar Quantities of Mixing

The word integral corresponds to the solution. Integral molar properties of solution mean the properties of the solution for its 1 g mole. Partial molar properties are related to the components of the solution and denote the properties of the components for their 1 g mole. Let us consider the formation of 1 g mole of solution AB by mixing $n_{A\,\text{moles}}$ of A and n_B moles of B. On dividing Eq. (6.9) by n_T (total number of moles in the solution), we obtain the following:

$$\frac{Q'}{n_T} = \overline{Q_A}\frac{n_A}{n_T} + \overline{Q_B}\frac{n_B}{n_T} = N_A\overline{Q_A} + N_B\overline{Q_B}$$

$$\text{or}\quad Q = N_A\overline{Q_A} + N_B\overline{Q_B} \tag{6.17}$$

where Q is the integral molar quantity of the solution. When the components are in their standard states (i.e. before mixing), then Eq. (6.17) can be written as:

$$Q^0 = N_A Q_A^0 + N_B Q_B^0$$

Change in the value of integral molar property (ΔQ_M) in the formation of 1 g mole of solution by mixing these components is given by the following expression:

$$\Delta Q_M = \left(N_A\overline{Q_A} + N_B\overline{Q_B}\right) - \left(N_A Q_A^0 + N_B Q_B^0\right)$$
$$\text{or}\quad \Delta Q_M = N_A\left(\overline{Q_A} - Q_A^0\right) + N_B\left(\overline{Q_B} - Q_B^0\right)$$
$$\text{or}\quad \Delta Q_M = N_A\Delta\overline{Q_A} + N_B\Delta\overline{Q_B} \tag{6.18}$$

where $\overline{Q_A}$ and $\overline{Q_B}$ are the partial molar properties of components A and B in the solution and Q_A^0 and Q_B^0 are the molar properties of components A and B in their standard (pure) states. In general, for a solution containing several components like 1,2,3, … n, the above equation is represented by the following expression:

$$\Delta Q_M = N_1\Delta\overline{Q_1} + N_2\Delta\overline{Q_2} + \cdots + N_n\Delta\overline{Q_n} \tag{6.19}$$

If the thermodynamic property is Gibbs free energy (G), then Eq. (xvii) is written as follows:

$$\Delta G_M = N_1\Delta\overline{G_1} + N_2\Delta\overline{G_2} + \cdots + N_n\Delta\overline{G_n}$$

By using Eq. (6.19), one can calculate the integral molar properties of a solution if the partial molar properties of its components are known.

6.6 Procedure to Obtain Partial Molar Quantities of Components in a Solution

From Eq. (6.15), the molar property 'Q' of the binary solution formed by mixing components A and B at constant temperature and pressure can be written as follows:

$$Q = N_A\overline{Q_A} + N_B\overline{Q_B}$$

where $\overline{Q_A}$ and $\overline{Q_B}$ are the partial molar properties (such as G, H and S) of components A and B in the solution. Differentiation of this equation yields the following:

$$\mathrm{d}Q = \overline{Q_A}\mathrm{d}N_A + \overline{Q_B}\mathrm{d}N_B + N_A\mathrm{d}\overline{Q_A} + N_B\mathrm{d}\overline{Q_B} \tag{6.20}$$

Now, Gibbs–Duhem equation gives $N_A\mathrm{d}\overline{Q_A} + N_B\mathrm{d}\overline{Q_B} = 0$. Hence, Eq. (6.18) becomes

$$\mathrm{d}Q = \overline{Q_A}\mathrm{d}N_A + \overline{Q_B}\mathrm{d}N_B \tag{6.21}$$

We know that $N_A + N_B = 1$ and $\mathrm{d}N_A = -\mathrm{d}N_B$. Multiplication of both sides of Eq. (6.21) by $\left(\frac{N_A}{\mathrm{d}N_B}\right)$ gives the following:

$$\left(\frac{N_A}{\mathrm{d}N_B}\right)\mathrm{d}Q = \overline{Q_A}\left(\frac{N_A}{\mathrm{d}N_B}\right)(-\mathrm{d}N_B) + \overline{Q_B}\left(\frac{N_A}{\mathrm{d}N_B}\right)\mathrm{d}N_B = -N_A\overline{Q_A} + N_A\overline{Q_B} \tag{6.22}$$

On addition of Eqs. (6.17) and (6.22), we get:

$$Q + \left(\frac{N_A}{\mathrm{d}N_B}\right)\mathrm{d}Q = N_A\overline{Q_A} + N_B\overline{Q_B} - N_A\overline{Q_A} + N_A\overline{Q_B}$$
$$= \overline{Q_B}(N_A + N_B) = \overline{Q_B}$$

or

$$\overline{Q_B} = Q + \left(\frac{N_A}{\mathrm{d}N_B}\right)\mathrm{d}Q = Q + N_A\left(\frac{\mathrm{d}Q}{\mathrm{d}N_B}\right) \tag{6.23}$$

Similarly, one can have

$$\overline{Q_A} = Q + \left(\frac{N_B}{\mathrm{d}N_A}\right)\mathrm{d}Q = Q + N_B\left(\frac{\mathrm{d}Q}{\mathrm{d}N_A}\right) \tag{6.24}$$

These are the expressions to correlate the partial molar properties of components of a binary solution with the molar property of the solution. Majority of the introductory thermodynamic texts are concerned with thermodynamic analysis of binary solutions because of their simplicity and can be extrapolated to study the thermodynamics of their respective ternary and multicomponent solutions. In terms of partial molar free energies of components A and B (i.e. $\overline{G_A}$ and $\overline{G_B}$), Eqs. (6.23) and (6.24) can be written as follows:

$$\overline{G_B} = G + N_A\left(\frac{\mathrm{d}G}{\mathrm{d}N_B}\right) \text{ and } \overline{G_A} = G + N_B\left(\frac{\mathrm{d}G}{\mathrm{d}N_A}\right)$$

6.7 Thermodynamics of Mixing of Components of Solutions

Free energy change in the formation of solution must be negative for complete mixing of components, otherwise formation of immiscible solution will occur. In other words, the sum of free energy of components after mixing must be less than the sum of free energies of these components before mixing. Higher negative values of free energy change in the dissolution process make the constituents more stable in the solutions. Formation of solution occurs either by evolution (exothermic) or absorption (endothermic) of heat energy, i.e. $\Delta H^M > 0$ or $\Delta H^M < 0$. The process of mixing changes the arrangement and order of individual components in the solution, and thus, the entropy change of mixing $\left(\Delta S^M\right)$ is always positive.

6.7.1 Free Energy Change in the Formation of Solution or Free Energy Change of Mixing

When 1 g-mole of another component is added to a large amount of solvent, the free energy change in the formation of solution is often referred to as the molar free energy of mixing and is, in general, denoted by the symbol ΔG^M. Superscript *M* indicates mixing of components. The net free energy change in the formation of solution is expressed as follows:

$$\Delta G^M = \sum G_{\text{Components}}(\text{solution}) - \sum G_{\text{Components}}(\text{pure})$$

Let components 1, 2, 3, ..., *i* are mixed together to form a solution. Changes in free energy of these components after mixing can now be expressed as follows:

$$\begin{aligned}
\Delta G_1^M &= N_1\left(\overline{G_1} - G_1^0\right) = RTN_1 \ln a_1 \\
\Delta G_2^M &= N_2\left(\overline{G_2} - G_2^0\right) = RTN_2 \ln a_2 \\
&\ldots \\
\Delta G_i^M &= N_i\left(\overline{G_i} - G_i^0\right) = RTN_i \ln a_i
\end{aligned}$$

Summation of these equations gives the net free energy change in the formation of a solution, and this expression can be written as follows:

$$\begin{aligned}
\Delta G^M &= N_1\left(\overline{G_1} - G_1^0\right) + N_2\left(\overline{G_2} - G_2^0\right) + \cdots + N_i\left(\overline{G_i} - G_i^0\right) \\
&= \left(N_1\overline{G_1} + N_2\overline{G_2} + \cdots + N_i\overline{G_i}\right) - \left(N_1G_1^0 + N_2G_2^0 + \cdots + N_iG_i^0\right) \\
\text{or } \Delta G^M &= RT(N_1 \ln a_1 + N_2 \ln a_2 + \cdots + N_i \ln a_i) \qquad (6.25)
\end{aligned}$$

For an ideal solution, $a_i = N_i$.

Hence,

$$\Delta G_{id}^M = RT(N_1 \ln N_1 + N_2 \ln N_2 + \cdots + N_i \ln N_i) \qquad (6.26)$$

Equations (6.25) and (6.26) are the required expressions of free energy change in the formation of real/actual and ideal solutions. It can be seen in these equations that for a fixed composition of the solution, the value of ΔG^M depends upon the temperature of solution formation. As temperature increases, the value of ΔG^M increases and vice versa.

6.7.2 Heat of Formation of a Solution or Heat of Mixing of the Components

Gibbs–Helmholtz Eq. (4.67) gives the following:

$$\left[\frac{\partial(G/T)}{\partial(1/T)}\right]_{P,\text{Comp.}} = H$$

For a component '*i*' in the pre-form, this equation takes the form as follows:

$$\left[\frac{\partial\left(G_i^0/T\right)}{\partial(1/T)}\right]_{P,\text{Comp.}} = H_i^0 \tag{6.27}$$

When the component '*i*' goes in solution, this equation can be written as follows:

$$\left[\frac{\partial\left(\overline{G_i}/T\right)}{\partial(1/T)}\right]_{P,\text{Comp.}} = \overline{H_i} \tag{6.28}$$

where H_i^0 and $\overline{H_i}$ are the partial molar enthalpies of component '*i*' in the standard state (i.e. before mixing) and in the solution. Subtraction of Eq. (6.27) from Eq. (6.28) gives the change in enthalpy of component '*i*' in forming a solution at constant pressure and composition and can be written as follows:

$$\overline{H_i} - H_i^0 = \left[\frac{\partial\left(\frac{\overline{G_i}-G_i^0}{T}\right)}{\partial\left(\frac{1}{T}\right)}\right]_{P,N_i}$$

or

$$\Delta H_i^M = \left[\frac{\partial\left(\Delta G_i^M/T\right)}{\partial(1/T)}\right]_{P,N_i} = \left[\frac{\partial(RT\ln a_i/T)}{\partial(1/T)}\right]_{P,N_i}$$

or

$$\Delta H_i^M = R\left[\frac{\partial \ln a_i}{\partial(1/T)}\right]_{P,N_i} \tag{6.29}$$

Use of Eq. (6.3), i.e. $a_i = \gamma_i N_i$, in the above equation leads to the following:

$$\Delta H_i^M = R\left[\frac{\partial \ln(\gamma_i N_i)}{\partial(1/T)}\right]_{P,N_i}$$

$$= R\left[\frac{\partial \ln \gamma_i)}{\partial(1/T)}\right]_{P,N_i} + R\left[\frac{\partial \ln N_i}{\partial(1/T)}\right]_{P,N_i}$$

or

$$\Delta H_i^M = R\left[\frac{\partial \ln \gamma_i)}{\partial(1/T)}\right]_{P,N_i} \quad \text{as} \quad \left[\frac{\partial \ln N_i}{\partial(1/T)}\right]_{P,N_i} = 0$$

This is the required equation for enthalpy change in component '*i*' on mixing. Similar equations can be obtained for other components involved in the formation of a solution, and the summation of all these equation gives the net enthalpy change (ΔH^M) in the formation a solution. Here, there are three possible cases as:

Case I: When the components in the solution exhibit positive deviation from ideality (i.e. $\gamma_i > 1$), then the value of ΔH^M becomes positive, indicating that the solution process is endothermic and ΔH^M is the quantity of heat absorbed per mole of the solution.

Case II: When the components of a solution show negative deviation from ideality (i.e. $\gamma_i < 1$), then the value of ΔH^M becomes negative and such solutions form exothermically and ΔH^M is the amount of heat liberated per mole of the solution formed at a temperature T.

Case III: When an ideal solution is formed, then $a_i = N_i$ and $\gamma_i = 1$. In this case, the value of ΔH^M becomes zero, i.e. $\Delta H_{id}^M = 0$. This means that the formation of an ideal solution is accompanied without any change in enthalpy, i.e. no any absorption or evolution of heat energy.

6.7.3 Entropy Change of Mixing or Entropy Change in the Formation of an Ideal Solution

From Eq. (4.22), we know that $\mathrm{d}G = -S\mathrm{d}T + V\mathrm{d}P$. At constant pressure and composition, this equation becomes

$$\mathrm{d}G = -S\mathrm{d}T \text{ or } -S = \mathrm{d}G/\mathrm{d}T \text{ or } -S = (\partial G/\partial T)_{P,\text{Comp.}}$$

In formation of a solution, the equation for entropy change can be written as follows:

$$-\Delta S^M = \left(\partial \Delta G^M/\partial T\right)_{P,\,\text{Comp}}$$

From Eq. (6.26), we know that for an ideal solution,

$$\Delta G_{id}^{M} = RT(N_1 \ln N_1 + N_2 \ln N_2 + \cdots + N_i \ln N_i)$$

Hence, the equation for ΔS^M can be expressed as follows:

$$\Delta S_{id}^{M} = -\left(\partial \Delta G_{id}^{M}/\partial T\right)_{P,\,\text{Comp.}} = -R(N_1 \ln N_1 + N_2 \ln N_2 + \cdots + N_i \ln N_i) \tag{6.30}$$

Alternative Method:

We know that $\Delta G^M = \Delta H^M - T\Delta S^M$. For an ideal solution, this can be written as follows:

$$\Delta G_{id}^{M} = \Delta H_{id}^{M} - T\Delta S_{id}^{M} \tag{6.31}$$

For an ideal solution, $\Delta H_{id}^{M} = 0$. Now Eq. (6.29) becomes $\Delta G_{id}^{M} = -T\Delta S_{id}^{M}$

$$\text{or}\quad \Delta S_{id}^{M} = -\Delta G_{id}^{M}/T = -RT(N_1 \ln N_1 + N_2 \ln N_2 + \cdots + N_i \ln N_i)/T$$
$$\text{or}\ \Delta S_{id}^{M} = -R(N_1 \ln N_1 + N_2 \ln N_2 + \cdots + N_i \ln N_i)$$

It is clear from this equation that the entropy change in the formation of an ideal solution is independent of the temperature of solution.

6.7.4 Volume Change in the Formation of an Ideal Solution

Equation (4.22) gives the following:

$$\mathrm{d}G = -S\mathrm{d}T + V\mathrm{d}P.$$

At constant temperature and composition, this equation becomes $\mathrm{d}G = V\mathrm{d}P$.

$$\text{or}\qquad V = \mathrm{d}G/\mathrm{d}P = (\partial G/\partial P)_{T,\,\text{Comp.}}$$

Let component '*i*' joins in the solution, then

$$\text{Volume of this component in the solution} = \overline{V}_i = \left(\partial \overline{G}_i/\partial P\right)_{T,\,\text{Comp.}} \tag{6.32}$$

$$\text{Volume of component 'i' before mixing} = V_i^0 = \left(\partial G_i^0/\partial P\right)_{T,\,\text{Comp.}} \tag{6.33}$$

Subtraction of Eq. (6.33) from Eq. (6.32) gives the equation for change in volume of component '*i*' after joining in the solution as follows:

$$\overline{V}_i - V_i^0 = \left[\frac{\partial(\overline{G}_i - G_i^0)}{\partial P}\right]_{T,\text{ Comp.}}$$

$$\text{or } \Delta V_i^M = \left(\frac{\partial \Delta G_i^M}{\partial P}\right)_{T,\text{Comp.}} \tag{6.34}$$

As outlined in Sect. 6.7.1, $\Delta G_i^M = RT \ln N_i$ for an ideal solution. Now, Eq. (6.34) can thus be written as follows:

$$\Delta V_i^{M,id} = RT(\partial \ln N_i/\partial P)_{T,\text{ Comp.}} = 0$$

This clarifies that the volume change in the component '*i*' accompanying the formation of an ideal solution is zero. Similarly, volume change in other components leading to the formation of an ideal solution can be proved to be zero. Hence,

Net change in volume of solution = sum of volume change in the components
= zero.

6.7.5 *Graphical Method for Determining Partial Molar Properties*

Let us consider a binary solution of components A and B. On the basis of Eqs. (6.25) and (6.30), we have the following:

$$\Delta G^M = RT(N_A \ln a_A + N_B \ln a_B) = N_A \Delta\overline{G}_A^M + N_B \Delta\overline{G}_B^M$$

$$\Delta S^{M,id} = -R(N_A \ln N_A + N_B \ln N_B) = N_A \Delta\overline{S}_A^M + N_B \Delta\overline{S}_B^M$$

Figures 6.2 and 6.3 present schematic representations for the variations of ΔG^M and $\Delta S^{M,id}$ with composition for a binary solution A–B at constant temperature. As shown in these figures, ΔG^M is negative and minimum at $N_A = N_B = 0.5$, and $\Delta S^{M,id}$ is positive and maximum at $N_A = N_B = 0.5$. $\Delta G^M = 0$ for pure A and B components, i.e. at $N_A = 1$ and $N_B = 1$, and this is at the top of the diagram. Figure 6.2 also shows the graphical procedure to evaluate the values of $\Delta\overline{G}_A^M$ and $\Delta\overline{G}_B^M$. In order to determine their values at any composition (say $N_A = N_B = 0.5$), a tangent is drawn at this point on the curve. Intercepts of the tangent with ΔG^M axis at $N_A = 1$ and $N_B = 1$ give the values of $\Delta\overline{G}_A^M$ (i.e. $RT \ln a_A$) and $\Delta\overline{G}_B^M$ (i.e.$RT \ln a_B$).

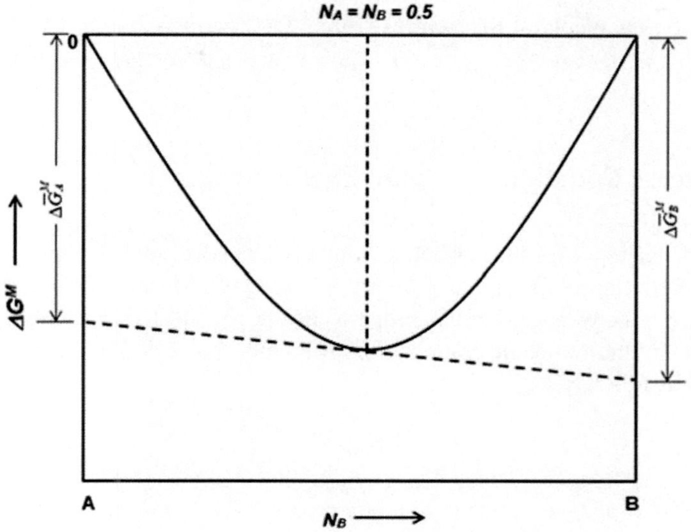

Fig. 6.2 ΔG^M versus composition curve in a binary solution A–B at constant temperature for graphical determination of the values of $\Delta\overline{G}_A^M$ and $\Delta\overline{G}_B^M$

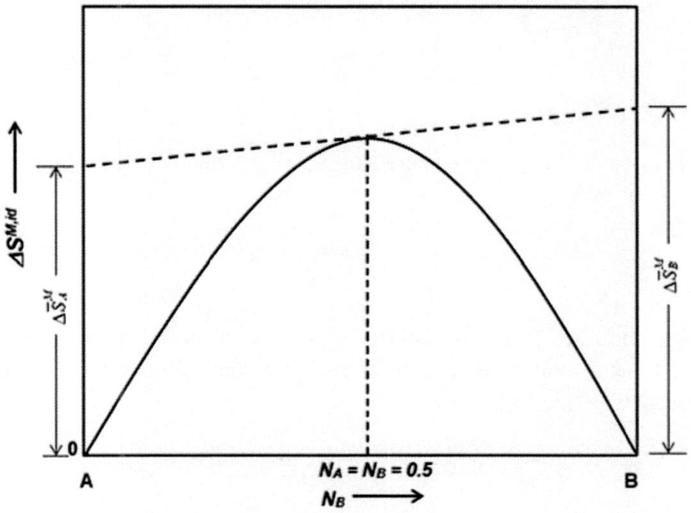

Fig. 6.3 $\Delta S^{M,id}$ versus composition curve in a binary solution A–B at constant temperature for graphical determination of the values of $\Delta\overline{S}_A^M$ and $\Delta\overline{S}_B^M$

Similarly, intercepts of the tangent with $\Delta S^{M,id}$ axis at $N_A = 1$ and $N_B = 1$ (Fig. 6.3) yield the values of $\Delta \overline{S}_A^M$ (i.e. $-RT \ln N_A$) and $\Delta \overline{S}_B^M$ (i.e. $-RT \ln N_B$).

6.8 Excess Function or Molar Excess Property

In case of dealing with real or actual solutions, an ideal of a function known as excess function or excess molar property has been introduced which is defined as the difference between the molar property in the actual or real solution and that in an ideal solution of same composition. It is only for extensive properties and is expressed as follows:

$$Z^{XS} = \Delta Z^{\text{real}} - \Delta Z^{\text{id}}$$

where ΔZ^{real} = change in extensive molar property Z ($G, H, S, \ldots.$) in the formation of a real solution, ΔZ^{id} = change in the value of the same property in the formation of an ideal solution at the same chemical composition and temperature and Z^{XS} = value of the same molar property Z in excess of that of an ideal solution.

Similarly, one can define the excess partial molar properties of a component 'i' in the solution as follows:

$$\overline{Z}_i^{XS} = \Delta \overline{Z}_i^{\text{real}} - \Delta \overline{Z}_i^{\text{id}}$$

The excess partial molar free energy of a component 'i' in the solution ($\overline{G}_i^{XS}$) is expressed as follows:

$$\overline{G}_i^{XS} = \Delta \overline{G}_i^{\text{real}} - \Delta \overline{G}_i^{\text{id}} \tag{6.35}$$

where $\Delta \overline{G}_i^{\text{real}}$ and $\Delta \overline{G}_i^{\text{id}}$ are the values of change in partial molar free energy of component 'i' as a result of its joining in real and ideal solutions. From Eqs. (6.25) and (6.26), we have the following:

$$\Delta \overline{G}_i^{\text{real}} = N_i RT \ln a_i \text{ and } \Delta \overline{G}_i^{\text{id}} = N_i RT \ln N_i$$

Now, Eq. (6.35) becomes as follows:

$$\overline{G}_i^{XS} = N_i RT \ln a_i - N_i RT \ln N_i = N_i RT \ln(a_i/N_i) = N_i RT \ln \gamma_i$$

The net excess free energy required in the formation of a real/actual solution can be expressed as follows:

$$G^{XS} = \Delta G^{M,\text{real}} - \Delta G^{M,\text{id}}$$

where $\Delta G^{M,\text{real}}$ and $\Delta G^{M,\text{id}}$ are the values of change in free energy in the formation of real and ideal solutions. Now, the above equation can be written as follows:

$$G^{XS} = \Delta G^{M,\text{real}} - \Delta G^{M,\text{id}} = \left(N_i RT \ln a_i + N_j RT \ln a_j + \ldots\right) - \left(N_i RT \ln N_i + N_j RT \ln N_j + \ldots\right)$$

where $i, j, \ldots$ are the components present in the solution and a_i, a_j, … and N_i, N_j, … are their activities and mole fractions in the solution. Now,

$$G^{XS} = N_i RT \ln(a_i/N_i) + N_j RT \ln\left(a_j/N_j\right) + \ldots$$

$$\text{or } G^{XS} = RT\left(N_i \ln \gamma_i + N_j \ln \gamma_j + \ldots\right) \quad (6.36)$$

This is the required equation of excess free energy in terms of activity coefficients of the components present in the solution. For an ideal solution, $\gamma_i = 1, \gamma_j = 1, \ldots$ Hence, $G^{XS,id} = 0$, i.e. formation of an ideal solution occurs without any requirement of excess free energy.

6.9 Regular Solutions

Hildebrand and coworkers first introduced the concept of regular solutions as being in between the ideal and real solutions. They defined a regular solution as one in which $\Delta S^M = \Delta S^M_{\text{id}}$ but $\Delta G^M = \Delta G^M_{\text{real}}$ and $\Delta H^M \neq 0$. This can also be written as follows:

$$\Delta S^M_{\text{reg}} = \Delta S^M_{\text{id}} = -R(N_1 \ln N_1 + N_2 \ln N_2 + \ldots) \text{ and}$$

$$\Delta G^M_{\text{reg}} = \Delta G^M_{\text{real}} = RT(N_1 \ln a_1 + N_2 \ln a_2 + \ldots)$$

This can be stated as "an entropy change in the formation of a regular solution is the same as that of an ideal solution of the same composition while the free energy change being equal to that of a real solution". We know that $\Delta G^M = \Delta H^M - T\Delta S^M$. Hence,

$$\Delta H^M_{\text{reg}} = \Delta G^M_{\text{reg}} + T\Delta S^M_{\text{reg}}$$

or

$$\Delta H^M_{\text{reg}} = RT(N_1 \ln a_1 + N_2 \ln a_2 + \ldots) - RT(N_1 \ln N_1 + N_2 \ln N_2 + \ldots)$$

or

$$\Delta H^M_{reg} = RT\left(N_1 \ln \frac{a_1}{N_1} + N_2 \ln \frac{a_2}{N_2} + \ldots\right)$$

As outlined in Sect. 6.2, $a_i = \gamma_i N_i$ where γ_i is the activity coefficient of component 'i' in the solution. Hence, the above equation can be expressed as follows:

$$\Delta H^M_{\text{reg}} = RT(N_1 \ln \gamma_1 + N_2 \ln \gamma_2 + \ldots) \tag{6.37}$$

where $\gamma_1, \gamma_2, \ldots$ are activity coefficients of components 1, 2,... in the solution. This is the required equation for enthalpy change in the formation of a regular solution.

By definition, $G^{XS} = H^{XS} - TS^{XS}$.

For a regular solution, $S^{XS} = \Delta S^{M,\text{real}} - \Delta S^{M,\text{id}}$ and $\Delta S^{M,\text{ real}} = \Delta S^{M,\text{ id}}$. Hence, $S^{XS} = 0$.

Now, $G^{XS} = H^{XS}$.

From Eq. (6.37), $\Delta H^M_{\text{reg}} = RT(N_1 \ln \gamma_1 + N_2 \ln \gamma_2 + \ldots)$.

Component-wise for a regular solution, this equation may be split as follows:

$$\Delta \overline{H}_1^{M,\text{reg}} = RT \ln \gamma_1, \; \Delta \overline{H}_2^{M,\text{reg}} = RT \ln \gamma_2, \ldots$$

In general,

$$\Delta \overline{H}_i^{M,\text{reg}} = RT \ln \gamma_i \tag{6.38}$$

6.10 Henry's Law

Henry's law is applicable to all solutions at extreme dilution, and its generalized form is $a_B = \gamma N_B$. Activity vs. mole fraction curves for a binary solution AB at a fixed temperature have been presented in Fig. 6.4. As can be seen in this figure, the variation of a_B is linear at very small values of N_B (i.e. when $N_B \to 0$), and the activity coefficient (γ) becomes constant and is represented by the symbol γ_B^0 within the concentration range of Henrian limit. In this composition range (usually termed as Henry's law region/Henrian limit), Henry's law is obeyed, i.e. the solute in binary dilute solution obeys this law. Henry's law for solute B within this region may be written as $a_B = \gamma_B^0 N_B$. Similarly, if $N_A \to 0$ (i.e. when component A becomes solute in a dilute solution), $a_A = \gamma_A^0 N_A$.

As is evident from Fig. 6.4, the plots for a_B join with Raoult's law line when $N_B \to 1$ (i.e. component B obeys Raoult's law on becoming solvent). On this observation, it may be specified that in the region where component B obeys Henry's law, component A tends to follow Raoult's law and vice versa. This can be proved as follows:

If the solute B in a binary solution AB obeys Henry's law, then $a_B = \gamma_B^0 N_B$

$$\text{or } \ln a_B = \ln \gamma_B^0 + \ln N_B$$

Differentiation of this equation yields $\mathrm{d} \ln a_B = \mathrm{d} \ln \gamma_B^0 + \mathrm{d} \ln N_B$.

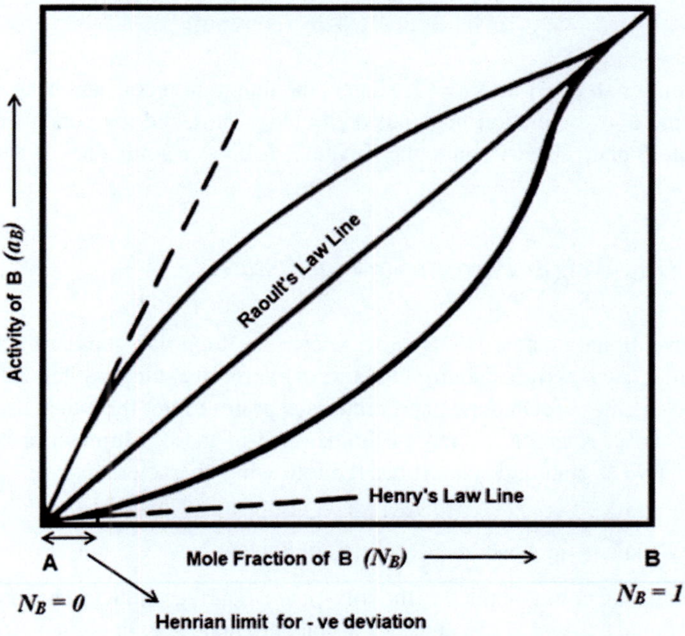

Fig. 6.4 Raoult's law and Henry's lines with positive and negative deviations from ideality

Now, $\mathrm{d}\ln\gamma_B^0 = 0$ because γ_B^0 is a constant. Hence, the above equation becomes as:

$$\mathrm{d}\ln a_B = \mathrm{d}\ln N_B \tag{6.39}$$

From Gibbs–Duhem Eq. (6.13), we know that $N_A\mathrm{d}\overline{G_A} + N_B\mathrm{d}\overline{G_B} = 0$

$$\text{or } RT(N_A\mathrm{d}\ln a_A + N_B\mathrm{d}\ln a_B) = 0$$

On insertion of Eq. (6.39) in this Gibbs–Duhem equation, we get the following:

$$\mathrm{d}\ln a_A = -\frac{N_B}{N_A}\mathrm{d}\ln N_B = -\frac{N_B}{N_A}\left(\frac{\mathrm{d}N_B}{N_B}\right) = -\frac{\mathrm{d}N_B}{N_A} \tag{6.40}$$

We know that $N_A + N_B = 1$. Hence, $\mathrm{d}N_A + \mathrm{d}N_B = 0$ or $\mathrm{d}N_A = -\mathrm{d}N_B$. Now, Eq. (6.40) becomes

$$\mathrm{d}\ln a_A = \frac{\mathrm{d}N_A}{N_A} = \mathrm{d}\ln N_A \tag{6.41}$$

Integration of this equation yields the following:

$$\ln a_A = \ln N_A + \ln \text{ (constant)}$$

By definition, $a_A = 1$ at $N_A = 1$. Hence, the integration constant becomes equal to 1, and the above equation becomes $a_A = N_A$. Thus, the net conclusion is that when solute B obeys Henry's law, the solvent A follows Raoult's law.

6.10.1 One Weight Percent Standard State

In the conventional standard state (also known as Raoultian standard state), pure solid/liquid is taken as the standard state. Use of alternative standard state is permitted in thermodynamics. Technocrats and industries prefer to get the calculation results directly in wt.%. Keeping in view of the demand of these, Chipman proposed the concept of 1 wt.% standard state. It has the following characteristics:

(i) Instead of mole fraction, the composition scale is in wt.%.
(ii) Henry's law is the basis and not the Raoult's law.

Let us consider components A as the solvent and B as the solute in a binary solution A–B. Henrian activity of B (h_B) in 1 wt.% standard state is expressed as follows:

$$h_B = \frac{a_B}{(a_B/W_B)_{W_B \to 0}} = (W_B)_{W_B \to 0}$$

where W_B is the wt.% of B. When $W_B \to 0$, Henry's law is obeyed. In other words, $h_B =$ wt.% B within Henrian limit. About 1 wt.% standard state means that when Henry's law region extends to 1 wt.% of B then $h_B = 1$. In case of positive deviation from Henry's law, a term known as Henrian activity coefficient (f_B) is introduced and is defined as follows:

$$f_B = h_B/(\text{wt.\%}B) \tag{6.42}$$

In Henry's law region, $f_B = 1$, and hence, $h_B =$ wt.% B.

6.11 Dilute Multicomponent Solutions—Interaction Between Solutes

In majority of the cases in Metallurgy and Materials Science, one has to often deal with multicomponent solutions. In multicomponent solutions, in addition to the interaction between solute and solvent atoms, a large number of interaction between the atoms of one solute with those of the other solutes occur and they have some effects on the activity coefficients of solutes. One important example of multicomponent solutions in Metallurgy is the liquid steel where Fe is the solvent and C, Si, Mn, S, P, etc. are present in very low concentration as solutes.

Scientist Wagner first proposed the concept of activity coefficient to quantify the solute–solute interactions. He derived an equation for activity coefficient of component '*i*' in a multicomponent solution (within Henrian limit):

$$\log f_i = \log f_i^0 + e_i^B(\text{wt.\%B}) + e_i^C(\text{wt.\%C}) + \ldots \tag{6.43}$$

where f_i = activity coefficient of component '*i*' in 1 wt.% standard state.

f_i^0 = Henry's law constant for 1 wt.% standard state.

$e_i^B, e_i^C, \ldots$ are known as interaction coefficients illustrating the influence of solutes B, C, ... on activity coefficient of component '*i*'.

Since, $f_i^0 = 1$, hence $\log f_i^0 = 0$.

Now, Eq. (6.43) becomes

$$\log f_i = e_i^B(\text{wt.\%B}) + e_i^C(\text{wt.\%C}) + \ldots \tag{6.44}$$

Solved Problems

Problem 6.1 In a Zn-Cd alloy, the activities of Zn and Cd at N_{Zn} (mole fraction of Zn) = 0.4 are 0.3 and 0.65 at a temperature of 227 °C. Find out the values of the followings at this temperature:

(a) Free energy change in the formation of ideal and real solutions $\left(\Delta G_{\text{id}}^M \text{ and } \Delta G_{\text{real}}^M\right)$.
(b) Excess free energy $\left(G^{XS}\right)$.
(c) Change in partial free energies $\left(\Delta \overline{G}_{\text{Zn}}^M, \Delta \overline{G}_{\text{Cd}}^M, \overline{G}_{\text{Zn}}^{XS} \text{ and } \overline{G}_{\text{Cd}}^{XS}\right)$ on mixing.
(d) Change in partial entropies $\left(\Delta \overline{S}_{\text{Zn}}^M \text{ and } \Delta \overline{S}_{\text{Cd}}^M\right)$ on mixing.
(e) Entropy and enthalpy changes in the formation of regular solution $\left(\Delta S_{\text{reg}}^M \text{ and } \Delta H_{\text{reg}}^M\right)$.

Solution

Data given: $N_{\text{Zn}} = 0.4$, $N_{\text{Cd}} = 0.6$, $a_{\text{Zn}} = 0.3$, $a_{\text{Cd}} = 0.65$ and $T = 500$ K.

(a) As per formula, $\Delta G_{id}^M = RT(N_{\text{Zn}} \ln N_{\text{Zn}} + N_{\text{Cd}} \ln N_{\text{Cd}})$

$$= 8.314 \times 500(0.4 \times \ln 0.4 + 0.6 \times \ln 0.6) = -2797.71\,\text{J/mol}$$

$$\Delta G_{\text{real}}^M = RT(N_{\text{Zn}} \ln a_{\text{Zn}} + N_{\text{Cd}} \ln a_{\text{Cd}}) = 8.314 \times 500(0.4 \times \ln 0.3 + 0.6 \times \ln 0.65)$$
$$= -3076.42\,\text{J/mol}$$

(b) As per formula, $G^{XS} = \Delta G^M_{real} - \Delta G^M_{id} = -3076.42 + 2797.71 = -278.71$ J/mol

(c) We know that, $\Delta \overline{G}_{Zn}^{M,\ real} = RT \ln a_{Zn} = 8.314 \times 500 \times \ln 0.3 = -5004.91$ J/mol

$\Delta \overline{G}_{Cd}^{M,real} = RT \ln a_{Cd} = 8.314 \times 500 \times \ln 0.65 = -1790.76$ J/mol
$\Delta \overline{G}_{Zn}^{M,id} = RT \ln N_{Zn} = 8.314 \times 500 \times \ln 0.4 = -3809.02$ J/mol
$\Delta \overline{G}_{Cd}^{M,id} = RT \ln N_{Cd} = 8.314 \times 500 \times \ln 0.6 = -2123.50$ J/mol
$\overline{G}_{Zn}^{XS} = \Delta \overline{G}_{Zn}^{M,real} - \Delta \overline{G}_{Zn}^{M,id} = -5004.91 + 3809.02 = -1195.89$ J/mol
$\overline{G}_{Cd}^{XS} = \Delta \overline{G}_{Cd}^{M,real} - \Delta \overline{G}_{Cd}^{M,id} = -1790.76 + 2123.50 = 332.74$ J/mol

(d) We know that, $\Delta \overline{S}_{Zn}^{M} = -R \ln N_{Zn} = 8.314 \times \ln 0.4 = -7.62$ J/K mol

$\Delta \overline{S}_{Cd}^{M} = -R \ln N_{Cd} = 8.314 \times \ln 0.6 = -4.25$ J/K mol

(e) We know that, $\Delta G^M = \Delta H^M - T\Delta S^M$

Hence, $\Delta G^M_{reg} = \Delta H^M_{reg} - T\Delta S^M_{reg}$ or $\Delta H^M_{reg} = \Delta G^M_{reg} + T\Delta S^M_{reg}$.

As known to us, $\Delta G^M_{reg} = \Delta G^M_{real}$ and $\Delta S^M_{reg} = \Delta S^M_{id}$. Thus, $\Delta H^M_{reg} = \Delta G^M_{red} + T\Delta S^M_{id}$

$\Delta S^M_{id} = -R(N_{Zn} \ln N_{Zn} + N_{Cd} \ln N_{Cd}) = -8.314 \times (0.4 \times \ln 0.4 + 0.6 \times \ln 0.6) = -5.6$ J/K mol $= \Delta S^M_{reg}$

Hence, $\Delta H^M_{reg} = -3076.42 + 500 \times (-5.6) = -5876.42$ J/mol

Problem 6.2 Following are the data for partial pressures of Zn in Cu–Zn alloy at 1333 K:

N_{Zn}	0.05	0.10	0.15	0.20	0.30	0.45	1.00
p_{Zn}(mm Hg)	22	45	90	180	456	970	3040

Answer the following:

(a) Does the alloy follow Raoult's law?
(b) Does Zn obey Henry's law? If yes, then in what composition range?
(c) Calculate the free energy change when 1 g atom of pure Zn at 1333 K dissolves in the alloy having $N_{Zn} = 0.3$.

Solution

We know that, $a_{Zn} = p_{Zn}/p^0_{Zn}$ and $\gamma_{Zn} = a_{Zn}/N_{Zn}$.

As per the given data, $p^0_{Zn} = 3040$ mm Hg.

The values of a_{Zn} and γ_{Zn} were calculated from the above formulae. The results are given in the following table:

N_{Zn}	0.05	0.10	0.15	0.20	0.30	0.45	1.00
a_{Zn}	0.00723	0.015	0.03	0.06	0.15	0.32	1.00
γ_{Zn}	0.145	0.15	0.20	0.30	0.50	0.71	1.00

(a) The value of γ_{Zn} is not one over the entire composition range of the solution. Hence, Raoult's law is not obeyed.
(b) The value of γ_{Zn} is nearly constant for N_{Zn} value < 0.15. Hence, Henry's law is obeyed in the concentration range 0.05–0.15.
(c) As per formula, $\Delta\overline{G}_{Zn}^{M} = RT \ln a_{Zn} = 8.314 \times 1333 \times \ln(0.15) = -$ 21,024.94 J/mol

Problem 6.3 For liquid Cu-Sn alloy at 1073 K, data given are as follows:

N_{Sn}	0.1	0.2	0.3	0.4	0.5	0.6	0.7	0.8	0.9
a_{Sn}	0.007	0.055	0.213	0.353	0.474	0.574	0.683	0.805	0.909
a_{Cu}	0.851	0.602	0.388	0.340	0.267	0.210	0.153	0.098	0.047

Answer the following:

(a) Do copper and tin follow Raoult's law?
(b) Do the components of the alloy obey Henry's law? If yes, then in what composition range?
(c) Calculate the values of $\Delta\overline{G}_{Sn}^{M}$, $\Delta\overline{G}_{Cu}^{M}$, $\Delta G^{M,\text{real}}$, G^{XS}, p_{Sn} and p_{Cu} for $N_{Sn} =$ 0.6.
(d) Whether a solution formed by mixing 60% Sn and 40% Cu is a regular solution?

Solution
By definition we know that activity coefficient, $\gamma_{Sn} = a_{Sn}/N_{Sn}$ and $\gamma_{Cu} = a_{Cu}/N_{Cu}$.
Calculated values of γ_{Sn} and γ_{Cu} are reported in the following table:

N_{Sn}	0.1	0.2	0.3	0.4	0.5	0.6	0.7	0.8	0.9
a_{Sn}	0.007	0.055	0.213	0.353	0.474	0.574	0.683	0.805	0.909
a_{Cu}	0.851	0.602	0.388	0.340	0.267	0.210	0.153	0.098	0.047
γ_{Sn}	0.07	0.275	0.71	0.883	0.948	0.957	0.976	1.01	1.01
γ_{Cu}	0.946	0.753	0.554	0.567	0.534	0.252	0.51	0.49	0.47

(a) The values of γ_{Sn} and γ_{Cu} are not 1 throughout the entire range of composition. Hence, Raoult's law is not obeyed.
(b) From the data listed in the above table, Henry's law is obeyed in the Sn concentration range of 0.8–0.9.

(c) As per formula, $\Delta\overline{G}_{\text{Sn}}^{M} = RT \ln a_{\text{Sn}} = 8.314 \times 1073 \times \ln(0.574) = -4952.23$ J/mol

$\Delta\overline{G}_{\text{Cu}}^{M} = RT \ln a_{\text{Cu}} = 8.314 \times 1073 \times \ln(0.210) = -13{,}922.42$ J/mol
We know that,

$$\begin{aligned}\Delta G^{M,\text{real}} &= RT(N_{\text{Sn}} \ln a_{\text{Sn}} + N_{\text{Cu}} \ln a_{\text{Cu}}) \\ &= 8.314 \times 1073[(0.6 \times \ln 0.574) + (0.4 \times \ln 0.210)] \\ &= -8540.3\,\text{J/mol}\end{aligned}$$

Now,

$$\begin{aligned}\Delta G^{M,\text{id}} &= RT(N_{\text{Sh}} \ln N_{\text{Sn}} + N_{\text{Cu}} \ln N_{\text{Cu}}) \\ &= 8.314 \times 1073[(0.6 \times \ln 0.6) + (0.4 \times \ln 0.4)] \\ &= -6003.88\,\text{J/mol}\end{aligned}$$

Now, $G^{XS} = \Delta G^{M,\text{ real}} - \Delta G^{M,\text{ id}} = -8540.3 + 6003.88 = -2536.42$ J/mol.
By using the formula, $\Delta\overline{G}_{\text{Sn}}^{M} = RT \ln p_{\text{Sn}}$ and considering $p_{\text{Sn}}^{0} = 1$ atm
or $\ln p_{\text{Sn}} = -4952.23/(8.341 \times 1073) = -0.56$.
or $p_{\text{Sn}} = 0.57$ atm.
Similarly, $p_{\text{Cu}} = 0.21$ atm.

(d) We know that for a regular solution, $S^{XS} = \Delta S^{M,\text{ real}} - \Delta S^{M,\text{id}} = 0$.

$\Delta S^{M,\text{real}} = -R(N_{\text{Sn}} \ln a_{\text{Sn}} + N_{\text{Cu}} \ln a_{\text{Cu}}) = -8.314 \times (0.6 \times \ln 0.574 + 0.4 \times \ln 0.21) = 7.96$ J/K mol
$\Delta S^{M,\text{real}} = -R(N_{\text{Sn}} \ln N_{\text{Sn}} + N_{\text{Cu}} \ln N_{\text{Cu}}) = -8.314 \times (0.6 \times \ln 0.6 + 0.4 \times \ln 0.4) = 5.60$ J/K mol
Now, $S^{XS} = 7.96 - 5.60 = 2.36$ J/K mol
Since, $S^{XS} \neq 0$, the solution having 60% Sn and 40% Cu is not a regular solution.

Problem 6.4 For liquid Cu-Ag alloy, the values of activity coefficient of Cu (γ_{Cu}) and Ag (γ_{Ag}) at a temperature of 1300 K and $N_{\text{Cu}} = 0.5$ are 1.36 and 1.32. Find out the values of the following:

$$\Delta\overline{G}_{\text{Cu}}^{M}, \Delta\overline{G}_{\text{Ag}}^{M}, \Delta G^{M,\text{ real}}, \Delta G^{M,\text{ id}}, G^{XS}, \alpha_{\text{Cu}}, \alpha_{\text{Ag}},$$
$$\Delta\overline{S}_{\text{Cu}}^{M}, \Delta\overline{S}_{\text{Ag}}^{M}, \Delta S^{M,\text{ real}}, \Delta S^{M,\text{ id}} \text{ and } S^{XS}.$$

Solution
As known to us, $\Delta\overline{G}_{\text{Cu}}^{M} = RT \ln a_{\text{Cu}}$, from the question, $\gamma_{\text{Cu}} = 1.36$ and $N_{\text{Cu}} = 0.5$

Hence, $a_{\text{Cu}} = 1.36 \times 0.5 = 0.68$

Now, $\Delta\overline{G}_{\text{Cu}}^{M} = 8.314 \times 1300 \times \ln 0.68 = -4168$ J/mol

As per definition, $\Delta\overline{G}^{M}_{\text{Ag}} = RT \ln a_{\text{Ag}}$.

As per question, $\gamma_{\text{Ag}} = 1.32$ and $N_{\text{Ag}} = 0.5$, hence $a_{\text{Ag}} = 1.32 \times 0.5 = 0.66$.

Now, $\Delta\overline{G}^{M}_{\text{Ag}} = 8.314 \times 1300 \times \ln 0.66 = -4490.97$ J/mol

We know that

$$\begin{aligned}\Delta G^{M,\text{ real}} &= RT(N_{\text{Cu}} \times \ln a_{\text{Cu}} + N_{\text{Ag}} \times \ln a_{\text{Ag}}).\\ &= 8.314 \times 1300(0.5 \times \ln 0.68 + 0.5 \times \ln 0.66) = -4329.64\,\text{J/mol}\end{aligned}$$

$$\begin{aligned}\Delta G^{M,\text{id}} &= RT\left(N_{\text{Cl}} \times \ln N_{\text{Cl}} + N_{\text{Ag}} \times \ln N_{\text{Ag}}\right)\\ &= 8.314 \times 1300(0.5 \times \ln 0.5 + 0.5 \times \ln 0.5) = -7491.67\,\text{J/mol}\\ G^{XS} &= \Delta G^{M,\text{ real}} - \Delta G^{M,\text{id}} = -4329.64 + 7491.67 = 3162\,\text{J/mol}\end{aligned}$$

From Eq. (6.15), we know that $\alpha_i = \frac{\ln \gamma_i}{(1-N_i)^2}$.
Hence, $\alpha_{\text{Cu}} = \frac{\ln \gamma_{\text{Cu}}}{(1-N_{\text{Cu}})^2} = \frac{\ln 1.36}{(0.5)^2} = 1.23$.
Similarly, $\alpha_{\text{Ag}} = \frac{\ln \gamma_{\text{Ag}}}{(1-N_{\text{Ag}})^2} = \frac{\ln 1.32}{(0.5)^2} = 1.11$.
As per definition, $\Delta\overline{S}^{M}_{\text{Cu}} = -R \ln a_{\text{Cu}} = -8.314 \times \ln 0.68 = 3.20\,\text{J/K mol}$.
For $N_{\text{Cu}} = 0.5$, $\Delta\overline{S}^{M}_{\text{Cu}} = 3.20 \times 0.5 = 1.6\,\text{J/K}$.
Similarly, $\Delta\overline{S}^{M}_{\text{Ag}} = -R \ln a_{\text{Ag}} = -8.314 \times \ln 0.66 = 3.45\,\text{J/K mol}$
For $N_{\text{Ag}} = 0.5$, $\Delta\overline{S}^{M}_{\text{Ag}} = 3.45 \times 0.5 = 1.725\,\text{J/K}$

$$\begin{aligned}\Delta S^{M,\text{real}} &= -R\left(N_{\text{Cu}} \ln a_{\text{Cl}} + N_{\text{Ag}} \ln a_{\text{Ag}}\right)\\ &= -8.314(0.5 \times \ln 0.68 + 0.5 \times \ln 0.66) = 3.330\,\text{J/K mol}\\ \Delta S^{M,\text{ id}} &= -R\left(N_{\text{Cu}} \ln N_{\text{Cu}} + N_{\text{Ag}} \ln N_{\text{Ag}}\right)\\ &= -8.314 \times (0.5 \ln 0.5 + 0.5 \ln 0.5) = 5.763\,\text{J/K mol}\\ S^{XS} &= \Delta S^{M,\text{ real}} - \Delta S^{M,\text{id}} = 3.330 - 5.763 = -2.433\,\text{J/K mol}\end{aligned}$$

Problem 6.5 Liquid Pb–Sn solution exhibits regular solution behaviour. At 746 K, the following relationship is given:

$$\ln \gamma_{\text{Pb}} = -0.737(1 - N_{\text{Pb}})^2$$

Determine the values of γ_{Sn}, γ_{Pb}, $\Delta\overline{H}^{M}_{\text{Pb}}$ and $\Delta\overline{H}^{M}_{\text{Sn}}$ in this alloy containing 60% Pb at the above-mentioned temperature.

Solution
From Eq. (6.15), we know that

$$\alpha_i = \frac{\ln \gamma_i}{(1 - N_i)^2} \quad \text{(P.6.1)}$$

Given equation is

$$\ln \gamma_{\mathrm{Pb}} = -0.737(1 - N_{\mathrm{Pb}})^2 \quad \text{(P.6.2)}$$

Comparison of Eq. (P.6.1) with Eq. (P.6.2) yields $\alpha_q = -0.73$.
From Eq. (6.16), we have the following:

$$\begin{aligned}
\ln \gamma_{\mathrm{Sn}} &= -N_{\mathrm{Sn}} N_{\mathrm{Pb}} \alpha_{\mathrm{Pb}} - \int_{N_{\mathrm{Sn}}=1}^{N_{\mathrm{Sn}}} \alpha_{\mathrm{Pb}} \mathrm{d}N_{\mathrm{Sn}} \\
&= -0.737 \times N_{\mathrm{Sn}} \times N_{\mathrm{Pb}} - \int_{1}^{N_{Sn}} (-0.737) \mathrm{d}N_{\mathrm{Sn}} \\
&= 0.737[N_{\mathrm{Sn}} N_{\mathrm{Pb}} + (N_{\mathrm{Sn}} - 1)] \\
&= 0.737[N_{\mathrm{Sn}}(1 - N_{\mathrm{Sn}}) - (1 - N_{\mathrm{Sn}})] \\
&= -0.737(1 - N_{\mathrm{Sn}})^2 \quad \text{(P.6.3)}
\end{aligned}$$

This is the required expression for γ_{Sn} at 746 K. For Pb–Sn alloy having 60% Pb, $N_{\mathrm{Sn}} = 0.4$.

From Eq. (P.6.3), $\ln \gamma_{\mathrm{Sn}} = -0.737(1 - N_{\mathrm{Sn}})^2 = -0.737(1 - 0.4)^2 = -0.737 \times 0.36 = -0.265$.

or $\gamma_{\mathrm{Sh}} = 0.77$.

$$\Delta \overline{H}^M_{\mathrm{Sn}} = RT \ln \gamma_{\mathrm{Sn}} = 8.314 \times 746 \times (-0.265) = -1643.59\mathrm{J/mol}$$

Similarly, from Eq. (P.6.2) $\ln \gamma_{\mathrm{Pb}} = -0.737(1 - N_{\mathrm{Pb}})^2 = -0.737(0.4)^2 = -0.118$

$$\Delta \overline{H}^M_{\mathrm{Pb}} = RT \ln \gamma_{\mathrm{Pb}} = 8.314 \times 746 \times (-0.118) = -731.86\mathrm{J/mol}$$

Problem 6.6 Find out the values of the partial molar enthalpies of mixing of Cd and Sn in a Cd-Sn alloy having 60% Cd. Data given at a temperature of 773 K are as follows:

N_{Cd}	0	0.1	0.3	0.5	0.7	0.9	1.0
ΔH^M (J/mol)	0	1247.7	2729.6	3347.2	2596.2	1052.3	0

Solution

A graph between ΔH^M and $N_{\mathrm{Cd}} = 0.6$ is plotted, and a tangent is drawn at $N_{\mathrm{Cd}} = 0.6$ (Fig. 6.5). The intercept of the tangent at $N_{\mathrm{Cd}} = 1.0$ gives the value of $\Delta \overline{H}^M_{\mathrm{Cd}}$ and is found to be 1649.7 J/mol. Similarly, the intercept at $N_{\mathrm{Cd}} = 0$ gives the value of $\Delta \overline{H}^M_{\mathrm{Sn}}$ and is found to be 5394.7 J/mol.

Problem 6.7 For a liquid Cu–Zn solution at 1400 K, it is given that $\alpha_{\mathrm{Zn}} = -\frac{19{,}250}{RT}$. Find out the following:

(a) An expression for γ_{Cu} at the above-mentioned temperature.
(b) Values of γ_{Cu}, γ_{Zn}, ΔG^M, $\Delta G^{M,\mathrm{id}}$ & G^{XS} at the above-mentioned temperature and $N_{\mathrm{Cu}} = 0.6$.

Solution

(a) We know that $\alpha_i = \frac{\ln \gamma_i}{(1-N_i)^2}$. Hence, for Cu–Zn solution $\ln \gamma_{\mathrm{Zn}} = \frac{-19{,}250}{RT}(1 - N_{\mathrm{Zn}})^2$.

For Cu, we can also derive the similar expression as follows:
From Eq. (6.16), we know that $\ln \gamma_{\mathrm{Cu}} = -N_{\mathrm{Cu}} N_{\mathrm{Zn}} \alpha_{\mathrm{Zn}} - \int_{N_{\mathrm{Cu}}=1}^{N_{\mathrm{Cu}}} \alpha_{\mathrm{Zn}} \mathrm{d}N_{\mathrm{Cu}}$

$$= \frac{19{,}250}{RT} N_{\mathrm{Cu}} N_{\mathrm{Zn}} + \int_{N_{\mathrm{Cu}}=1}^{N_{\mathrm{Cu}}} \frac{19{,}250}{RT} \mathrm{d}N_{\mathrm{Cu}} = \frac{19{,}250}{RT} [N_{\mathrm{Cu}} N_{\mathrm{Zn}} + (N_{\mathrm{Cu}} - 1)]$$

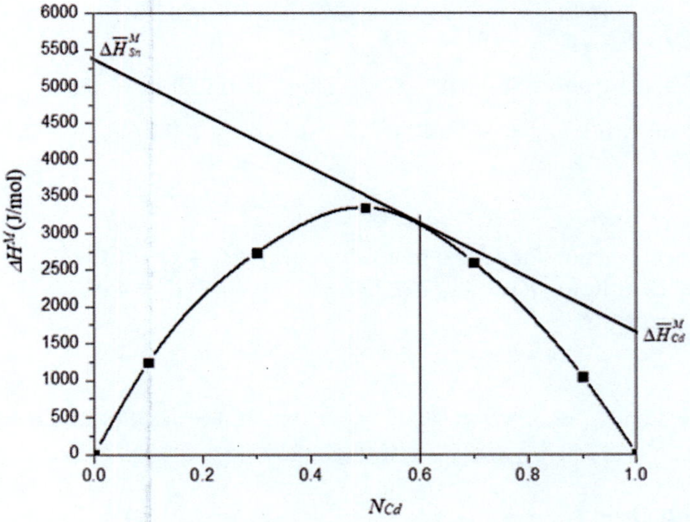

Fig. 6.5 Plot between ΔH^M and $N_{\mathrm{Cd}} = 0.6$ as per the data given in the problem

$$= \frac{19{,}250}{RT}[N_{Cu}(1 - N_{Cu}) + N_{Cu} - 1] = \frac{19{,}250}{RT}\left[2N_{Cu} - N_{Cu}^2 - 1\right]$$
$$= -\frac{19{,}250}{RT}(1 - N_{Cu})^2$$

Now, $\ln\gamma_{Cu} = -\frac{19{,}250}{RT}(1 - N_{Cu})^2$ is the required expression for γ_{Cu}.

(b) At $N_{Cu} = 0.6$, $\ln\gamma_{Cu} = -\frac{19{,}250}{8.314\times1400}(1 - 0.6)^2 = -0.265$ or $\gamma_{Cu} = 0.77$.

Now, $\ln\gamma_{Zn} = \frac{-19{,}250}{RT}(1 - N_{Zn})^2 = \frac{-19{,}250}{8.314\times1400}(1 - 0.4)^2 = -0.595$ or $\gamma_{Zn} = 0.55$.

We know that

$$\begin{aligned}\Delta G^{M,\text{red}} &= RT(N_{Cu} \times \ln a_{Cl} + N_{Zn} \times \ln a_{Zn}) \\ &= 8.314 \times 1400 \times [0.6 \times \ln(0.77 \times 0.6) + 0.4 \times \ln(0.55 \times 0.4)] \\ &= -12{,}442.33\ \text{J/mol}\end{aligned}$$

$$\begin{aligned}\Delta G^{M,id} &= RT(N_{Cu} \times \ln N_{Cu} + N_{Zn} \times \ln N_{Zn}) \\ &= 8.314 \times 1400 \times [0.6 \times \ln 0.6 + 0.4 \times \ln 0.4] = -7833.59\ \text{J/mol}\end{aligned}$$

Now, $G^{XS} = \Delta G^{M,\text{ real}} - \Delta G^{M,\text{id}} = -12{,}442.33 + 7833.59 = -4608.74\ \text{J/mol}$.

Problem 6.8 Calculate the wt.% of oxygen in liquid steel for the removal of carbon from hot metal having 3 wt.% C, 0.8 wt.% Mn, 1.2% wt.Si, 0.4 wt.% P and 0.04 wt.% S at a temperature of 1873 K. Data given are as follows:

$$\log K = \frac{1056}{T} + 2.131$$
$$e_C^C = 0.14,\ e_C^{Mh} = -0.012,\ e_C^{Si} = 0.08,\ e_C^P = 0.051,\ e_C^S = 0.046,$$
$$e_O^C = -0.13,\ e_O^{Mn} = -0.021,\ e_O^{Si} = -0.131,\ e_O^P = 0.07,\ e_O^S = -0.133$$

Solution

The chemical reaction for carbon removal in steel is $C + \underline{O} = CO$.

At $T = 1873$ K, $\log K = \frac{1056}{1873} + 2.131 = 2.69$ or $K = 489.78$.

We know that. $K = \frac{p_{CO}}{h_C \times h_O}$.

Since,

$$p_{CO} = 1 \text{ and } h_i = f_i \times (\text{wt.}\%i),\quad f_C \times (\text{wt.}\%\text{C}) \times f_O \times (\text{wt.}\%\text{O}) = 1/K \tag{P.6.4}$$

From Eq. (6.44),

$$\log f_C = e_C^S(\text{wt.\%S}) + e_C^C(\text{wt.\%C}) + e_C^{Si}(\text{wt.\%Si}) + e_C^{Mn}(\text{wt.\%Mn}) + e_C^P(\text{wt.\%P})$$
$$= 0.046 \times 0.04 + 0.14 \times 3.0 + 0.08 \times 1.2 - 0.012 \times 0.8 + 0.051 \times 0.4$$
$$= 0.5286$$

or

$$f_C = 3.38$$

Similarly,

$$\log f_O = e_O^S(w_t\%S) + e_O^C(\text{wt.\%C}) + e_O^{Si}(\text{wt.\%Si}) + e_O^{Mn}(\text{wt.\%Mn}) + e_O^P(\text{wt.\%P})$$
$$= -0.133 \times 0.04 - 0.13 \times 3.0 - 0.131 \times 1.2 - 0.021 \times 0.8 + 0.07 \times 0.4$$
$$= -0.5413$$

or

$$f_O = 0.29$$

Now, on inserting these values in Eq. (P.6.4), we get 3.38 × 3.0 × 0.29 × (wt.% O) = 1/489.78.

$$\text{or} \quad \text{wt.\% O} = 6.94 \times 10^{-4}$$

Problem 6.9 Calculate the value of activity (in 1 wt.% standard state) of sulphur in hot metal obtained from a blast furnace at a temperature of 1873 K. Data given are as follows:

Chemical composition (in wt.%) of hot metal: C = 4.0%, S = 0.04%, Si = 2.0% and Mn = 1.0%

$$e_S^C = 0.24,\ e_S^{Mh} = -0.025,\ e_S^{SS} = 0.066,\ e_S^S = -0.028$$

Solution
We know that $h_S = f_S \times$ (wt.% S)

and
$$\log f_S = e_S^S(\text{wt}_u\%S) + e_S^C(\text{wt.\%C}) + e_S^{Si}(\text{wt.\%Si}) + e_S^{Mn}(\text{wt.\%Mn})$$
$$= -0.028 \times 0.04 + 0.24 \times 4.0 + 0.066 \times 2.0 - 0.025 \times 1.0 = 1.066.$$

or

$$f_S = 11.64.$$

Hence, $h_S = 11.64 \times 0.04 = 0.465$.

Exercise

6.1 The values of Activity of Sn and Cu in a liquid Cu-Sn alloy at a temperature of 1400 K and $N_{Sn} = 0.4$ are 0.333 and 0.362. Find out the values of the following:

p_{Sn}, p_{Cu}, γ_{Sn}, γ_{Cu}, $\Delta\overline{G}^M_{Sn}$, $\Delta\overline{G}^M_{Cu}$, $\Delta G^{M,real}$ and G^{XS}.

[**Ans**: $p_{Sn} = 6.3 \times 10^{-7}$atm, $p_{Cu} = 3.68 \times 10^{-7}$atm, $\gamma_{Sn} = 0.833$, $\gamma_{Cu} = 0.603$, $\Delta\overline{G}^M_{Sn} = -12{,}800$ J/mol, $\Delta\overline{G}^M_{Cu} = -13{,}046$ J/mol, $\Delta G^{M,real} = -12{,}950$ J/mol, $G^{XS} = -5117$ J/mol]

6.2 The values of activity of Al at its different mole fraction values in Al-Cu solution at 1373 K are given in the following table:

N_{Al}	0.10	0.20	0.30	0.38	0.45	0.53	0.63	0.75	0.90
a_{Al}	0.008	0.005	0.03	0.10	0.20	0.31	0.50	0.69	0.89

Find out the value of activity and activity coefficient of Al in a solution containing 49% Al.

[**Ans**: at $N_{Al} = 0.49$, $a_{Al} = 0.25$ and $\gamma_{Al} = 0.51$]

6.3 The values of free energy change (ΔG^M in kJ/mol) involved in the formation of liquid Sn–Cu alloys at 1593 K as a function of composition have been listed in the table given below. With the help of these data, find out the values of activity of Cu and Sn at the above temperature and $N_{Sn} = 0.5$.

N_{Sn}	0	0.1	0.2	0.3	0.4	0.5	0.6	0.7	0.8	0.9	1.0
ΔG^M	0	7.43	11.45	13.32	13.89	13.61	12.64	10.94	8.46	5.28	0

[*Hint: First, find out the values of* $\Delta\overline{G}^M_{Cu}$ *and* $\Delta\overline{G}^M_{Sn}$ *from* ΔG^M *versus* N_{Sn} *plot and then use the formula* $\Delta\overline{G}^M_i = RT \ln a_i$]

6.4 Calculate the value of activity of sulphur in liquid steel in 1 wt.% standard state at 1873 K. Following data are given:

Chemical composition (in wt.%) of liquid steel: C = 0.8%, S = 0.04%, Si = 0.1%, P = 0.04% and Mn = 0.5%.

$e^C_S = 0.11$, $e^{Mn}_S = -0.026$, $e^{Si}_S = -0.063$, $e^S_S = -0.028$, $e^P_S = 0.029$.

[**Ans**: $h_S = 0.048$]

6.5 Deoxidation of molten steel by aluminium occurs at a temperature of 1873 K. The molten steel in a bath contains 0.1 wt.% C and 1.0 wt.% Mn. The final oxygen content in this steel is to be reduced to 0.001 wt.%. Find out the residual Al content of molten steel. Data given are as follows:

$$\log k_{Al} = \frac{-58{,}473}{T} + 17.74$$

$e_{\mathrm{Al}}^{\mathrm{Mn}} = 0$, $e_{\mathrm{Al}}^{\mathrm{C}} = 0.091$, $e_{\mathrm{Al}}^{\mathrm{O}} = -6.6$, $e_{\mathrm{Al}}^{\mathrm{Al}} = 0$, $e_{\mathrm{O}}^{\mathrm{Mn}} = -0.021$, $e_{\mathrm{O}}^{\mathrm{C}} = -0.45$, $e_{\mathrm{O}}^{\mathrm{O}} = -0.2$, $e_{\mathrm{O}}^{\mathrm{Al}} = 0$.

[**Ans:** 7.193×10^{-3} wt.%]

6.6 For liquid Fe–Mn alloys at 1590 °C, the values of (in J/mol) with composition have been listed in the table given below:

N_{Mn}	0.9	0.8	0.7	0.6	0.5	0.4	0.3	0.2	0.1
G^{XS}	395	703	925	1054	1100	1054	925	703	395

Calculate the values of the following:

(a) $\overline{G}_{\mathrm{Fe}}^{XS}$ and $\overline{G}_{\mathrm{Mn}}^{XS}$ at $N_{\mathrm{Mn}} = 0.6$
(b) ΔG^{M} at $N_{\mathrm{Mn}} = 0.4$

[**Ans:** $\overline{G}_{\mathrm{Fe}}^{XS} = 1583$ J/mol and $\overline{G}_{\mathrm{Mn}}^{XS} = 703$ J/mol; $\Delta G^{\mathrm{M}} = -9.37$ J/mol]

Appendix A

Thermodynamic Constants

1. Absolute zero, 0 K = − 273.14 °C
2. Avogadro's number, $N_a = 6.0228 \times 10^{23}$ mol^{-1}
3. Boltzmann's constant, $k = 1.38047 \times 10^{-16}$ erg/mol K
4. Gram-molar volume of gas at standard temperature and pressure (i.e. 0 °C and 1 atm pressure) = 22.414 L
5. Electronic charge,

$$
\begin{aligned}
e &= 1.60203 \times 10^{-20}\,\text{abs. e.m.u.}\\
&= 4.8024 \times 10^{-10}\,\text{e.s.u.}\\
&= 1.602 \times 10^{-19}\,\text{coulombs}
\end{aligned}
$$

6. Electronic mass, $em = 9.1072 \times 10^{-28}$ gm
7. Ice point, 0 °C = 273.14 K
8. Plank's constant, $h = 6.6242 \times 10^{-27}$ erg.s
9. Universal gas constant,

$$
\begin{aligned}
R &= 1.9872\,\text{cal/molK}\\
&= 8.314\,\text{J/K mol}\\
&= 0.0821\,\text{L(atm)/K mol}
\end{aligned}
$$

10. Velocity of light, $v = 2.9998 \times 10^{10}$ cm/s
11. 1 abs. joule = 0.02389 cal. $= 10^7$ erg
12. 1 international joule $= 1.0002 \times 10^7$ erg $= 1.0002$ abs.joule
13. 1 calorie = 4.185 abs. joule
14. 1 Bar $= 10^5\,\text{N/m}^2 = 10^6\,\text{dynes/cm}^2$
15. 1 Faraday = 96,487 abs. coulombs/gm − equivalent = 23,052 cal/volt

S. K. Sahoo et al., *Fundamentals of Metallurgical Thermodynamics*,
https://doi.org/10.1007/978-981-99-6671-4

16. 1 electron-volt = 1.602×10^{-19} J
17. 1 litre = 1000.028 c.c.
18. 1 litre-atm = 1.013×10^{9} erg

Appendix B

Metals and Some of their Physical Properties

Metal	Atomic number	Atomic weight	Density (gm/cm^3)	Melting point (°C)
Al	13	26.98	2.7	660
Ag	47	107.873	10.49	960.5
As	33	74.91	5.727	814
Au	79	197.2	19.32	1063
B	5	10.82	2.34	2300
Ba	S6	137.36	3.5	710
Be	4	9.013	1.845	1284
Bi	83	209.0	9.8	271.3
Ca	20	40.08	1.54	851
Cb	41	92.91	9.57	2468
Cd	48	112.41	8.65	320.9
Ce	58	140.13	6.66	795
Co	27	58.94	8.90	1493
Cr	24	52.01	7.19	1875
Cs	55	132.91	1.873	28.5
Cu	29	63.54	8.94	1083
Dy	66	162.51	8.536	1407
Er	68	167.21	9.051	1497
Eu	63	152.0	5.259	826
Fe	26	55.85	7.87	1535
Ga	31	69.72	5.907	29.75
Gd	64	157.26	7.895	1312

(continued)

S. K. Sahoo et al., *Fundamentals of Metallurgical Thermodynamics*,
https://doi.org/10.1007/978-981-99-6671-4

(continued)

Metal	Atomic number	Atomic weight	Density (gm/cm^3)	Melting point (°C)
Ge	32	72.60	5.32	936
Hf	72	178.50	13.29	2150
Hg	80	200.61	13.55	38.87
Ho	67	164.94	8.803	1461
In	49	114.82	7.31	156.6
Ir	77	192.2	22.42	2410
K	19	39.10	0.87	63.7
La	57	138.92	6.174	920
Li	3	6.940	0.534	179
Lu	71	174.99	9.842	1652
Mg	12	24.32	1.74	651
Mn	25	54.94	7.44	1244
Mo	42	95.95	10.22	2610
Na	11	22.991	0.97	97.9
Nd	60	144.27	7.004	1024
Ni	28	58.69	8.9	1452
Os	76	190.2	22.5	3000
Pb	82	207.21	11.34	327.4
Pd	46	106.4	12.02	1552
Pt	78	195.09	21.40	1769
Pu	94	239.11	19.84	639.5
Pr	59	140.92	6.782	935
Pm	61	145		1035
Re	75	186.22	21.02	3180
Rh	45	102.91	12.44	1960
Ru	44	101.1	12.4	2250
Rb	37	85.48	1.53	38.5
Sb	51	121.76	6.68	630.5
Sc	21	44.96	2.99	1539
Se	34	78.96	479	217
Si	14	28,09	233	1410
Sm	62	150.35	7.536	1072
Sn	50	118.7	7.3	232
Sr	38	87.63	2.6	770
Ta	73	180.95	16.6	2996
Te	S2	127.61	6.25	449.5

(continued)

(continued)

Metal	Atomic number	Atomic weight	Density (gm/cm^3)	Melting point (°C)
Tb	65	158.93	8.272	1356
TI	81	204.39	11.85	303
Th	90	232.05	11.66	1750
Tm	69	168.94	9.332	1545
Ti	22	47.90	4.54	1668
W	74	183.92	19.3	3410
U	92	238.07	19.07	1132
Yb	70	173.04	6.977	824
Y	39	88.92	4.472	1509
Zn	30	65.38	7.133	419.5
Zr	40	91.22	6.45	1852

Appendix C

Some Other Physical Properties of Metals

Metal	Boiling point (°C)	Latent heat of fusion (cal/gm)	Latent heat of vaporization (cal/gm)	Specific heat (cal/gm)
Al	2450	96	3050	0.214 (20 °C)
Ag	2212	25	556	0056 (20 °C)
As	615	88.5	102	0.078
Au	2966	14.96	415	0.0312 (18 °C)
B	2507	250–275	5917	0.425 (20 °C)
Ba	1500	13.3	262	0.068 (20 °C)
Be	1627	12.5	204.3	0.0294 (20 °C)
Bi	2550	489	8300	0.307 (25 °C)
Ca	1482	55.7	918	0.149 (20 °C)
Cb	4927	69	1782	0.0642 (20 °C)
Cd	767	13.2	286.4	0.055 (28 °C)
Ce	3468	15.7	679	0.049 (20 °C)
Co	3100	62	1500	0.1056 (20 °C)
Cr	2199	61.5	1474	0.1068 (20 °C)
Cs	705	313.76	146	0.052 (20 °C)
Cu	2595	48.9	1150	0.092 (20 °C)
Dy	2600	25.2	412	0.0413 (0 °C)
Er	2900	24.5	401	0.0398 (0 °C)
Eu	1439	15.15	276	0.0395 (0 °C)
Fe	300	66.2	1515	0.107 (20 °C)

(continued)

S. K. Sahoo et al., *Fundamentals of Metallurgical Thermodynamics*,
https://doi.org/10.1007/978-981-99-6671-4

(continued)

Metal	Boiling point (°C)	Latent heat of fusion (cal/gm)	Latent heat of vaporization (cal/gm)	Specific heat (cal/gm)
Ga	1983	19.16	1014	0.0977 (29 °C)
Gd	3000	23.6	459	0.0713 (0 °C)
Ge	2700	111.5	1200	0.086 (25 °C)
Hf	5400	29.1	885	0.0352 (20 °C)
Hg	357	2.8	69.7	0.03325 (20 °C)
Ho	2600	24.8	405	0.0391 (0 °C)
In	2075	6.8	484	0.058 (20 °C)
Ir	5300	32.6	790	0.032 (20 °C)
K	760	14.6	496	0.18 (14 °C)
Li	1317	158	4680	0.784 (0 °C)
Lu	3327	26.3	515	0.0368 (0 °C)
Mg	1103	82.2	1337	0.25 (25 °C)
Mn	2097	63.7	977.6	0.114 (20 °C)
Mo	5560	69.8	1222	0.066 (0 °C)
Na	883	27.05	1005	0.295 (20 °C)
Nd	3027	18.0	179	0.0499 (0 °C)
Ni	2900	73.8	1487	0.105 (20 °C)
Os	5500	36.9	790	0.039 (20 °C)
Pb	1737	5.89	204.6	0.031 (20 °C)
Pd	3980	37.8	88.3	0.0584 (20 °C)
Pt	4530	24.1	625	0.0314 (20 °C)
Pu	3235	3	336.6	0.034 (25 °C)
Pr	3127	17.0	560	0.0458 (0 °C)
Pm	2730		415	
Rb	688	6.1	212	0.080 (0 °C)
Re	5900	42.4	815	0.03262 (20 °C)
Rh	4500	50.5	1150	0.059 (20 °C)
Ru	4900	60.3	1340	0.057 (20 °C)
Sb	1440	38.3	383	0.0504 (20 °C)
Sc	2727	85.3	1743	0.1332 (20 °C)
Se	685	16.5	79.6	0.081 (20 °C)
Si	2480	395	2530	9.1597 (0 °C)
Sm	1900	17.3	306	0.0431 (20 °C)
Sn	2270	14.5	573	00,542 (20 °C)
Sr	1380	25	383	0.176 (20 °C)

(continued)

(continued)

Metal	Boiling point (°C)	Latent heat of fusion (cal/gm)	Latent heat of vaporization (cal/gm)	Specific heat (cal/gm)
Ta	6100	41.5	995	0.034 (0 °C)
Tb	2800	24.5	440	0.041 (0 °C)
Te	990	32	95	0.047 (20 °C)
Th	4200	19.8	560	0.0282 (20 °C)
Ti	3260	104.5	2350	0.125 (20 °C)
Tl	1457	5.04	189.9	0.031 (20 °C)
Tm	1727	26.0	348	0.0381 (0 °C)
W	5900	46	1038	0.032 (20 °C)
U	3813	19.75	420	0.028 (20 °C)
V	3000	82.5	2150	0.120 (0 °C)
Yb	1427	12.71	220	0.0347 (0 °C)
Y	3200	46.2	1045.8	0.074 (50 °C)
Zn	910	24.4	419.5	0.0925 (20 °C)
Zr	3580	60.2	1360	0.0659 (20 °C)

References

Bodsworth C (2019) Physical chemistry of iron and steel manufacture. CBS Publishers and Distributors Pvt. Ltd., New Delhi
Darken LS, Gurry RW (1953) Physical chemistry of metals. McGraw Hill, New York
Dutta SK, Lele AB (2012) Metallurgical thermodynamics, kinetics and numericals. S. Chand and Company Ltd., New Delhi
Gaskell DR (1973) Introduction to metallurgical thermodynamics. McGraw Hill, New York
Ghosh A (2003) Text book of materials and metallurgical thermodynamics. Prentice Hall of India, New Delhi
Kapoor ML (1981) Chemical and metallurgical thermodynamics. NEM Chand and Bros, Roorkee
Ray HS, Sridhar R, Abraham KP (2008) Extraction of non-ferrous metals. East-West Press Pvt. Ltd., New Delhi
Shamsuddin M (2016) Physical chemistry of metallurgical processes. Willey, Hoboken
Tupkary RH (1995) Introduction to metallurgical thermodynamics. Tu Publishers, Nagpur
Upadhyaya GS, Dube RK (1977) Problems in metallurgical thermodynamics and kinetics. Pergamon Press, Oxford

S. K. Sahoo et al., *Fundamentals of Metallurgical Thermodynamics*,
https://doi.org/10.1007/978-981-99-6671-4

Index

A

Activity, 3, 12, 28, 94, 125, 130–135, 149, 167, 169–172, 174, 175, 185, 186, 189, 197, 198

Activity coefficient, 133, 169, 170, 175, 185, 186, 188, 189, 191, 192, 198

Activity in binary solutions, 186

Activity quotient, 132, 134

Adiabatic process, 7, 14, 20, 21, 36, 37, 52, 53, 59

Allotropic transformation in pure iron, 94

Alpha function, 175

Association of molecules and entropy change, 72

B

Binary solutions, 168, 170, 171, 175–177, 182, 183, 186, 188

Boiling temperature, 109

C

Carbothermic reduction, 139, 147, 152

Carnot cycle, 49, 51, 52, 56, 66

Chemical potential, 13, 131, 132, 173

Chemical reaction and enthalpy change, 15

Chemical reaction and entropy change, 69

Chemical reaction and free energy change, 99, 116, 134

Chemical thermodynamics, 2, 3, 12, 17, 84, 130

Classical thermodynamics, 1–3, 12

Clausius-Clapeyron equations, 81, 105–109, 118, 119, 121, 123

Closed system, 5, 19, 102, 103

Cold mechanical working and entropy change, 71

Combined statements of first and second laws of thermodynamics, 81, 84

Compression, 7–9, 11, 20, 48, 49, 51, 57

Constant pressure processes, 23

Constant volume processes, 22

Criteria for establishment of equilibrium, 11, 87, 88

Cyclic processes, 16, 27, 63

D

Decomposition of metal oxides and free energy change, 139, 153

Decomposition temperature with fugacity and pressure, 128

Degradation of energy, 47, 68

Deoxidation of steel, 125, 149, 150

Dissociation of molecules and entropy change, 72

Dissociation of sulphides, 152

E

Efficiency of heat engines, 48, 49

Ellingham diagrams, 1, 125, 139–141, 143, 145, 147–154

Endothermic, 20, 21, 29, 62, 93, 137, 143, 146, 171, 178, 180

Enthalpy, 1, 4, 10, 12, 15, 20, 21, 23, 27–30, 32, 34–36, 61, 82, 88, 92, 95, 96, 99, 103, 112, 116, 117, 135, 145, 146, 167, 173, 179, 180

Enthalpy change during phase transformation, 15, 30, 32, 33

S. K. Sahoo et al., *Fundamentals of Metallurgical Thermodynamics*, https://doi.org/10.1007/978-981-99-6671-4

Enthalpy changes, 15, 20, 27–34, 36–38, 40, 42, 43, 45, 46, 62, 77–80, 88, 106, 146, 153, 155, 180, 186, 189
Enthalpy change with change of state, 15
Enthalpy change without change of state, 15
Entropy, 1–4, 10, 12, 47, 48, 50, 55, 56–63, 67–72, 76, 79, 82, 83, 86–88, 95, 96, 110–117, 120, 143, 146, 148, 154, 167, 171, 173, 174, 189
Entropy change during melting, 62
Entropy change during vaporization, 62
Entropy changes, 27, 47, 56–58, 61–63, 68, 69, 71–74, 76–79, 96, 109, 113–115, 117, 136, 143, 145, 146, 153, 155, 178, 180, 181, 185
Equation of state, 105, 126
Equations for entropy change under isochoric and isobaric conditions, 47
Equilibrium, 1–3, 7, 10–13, 47, 48, 56, 57, 63, 66, 68–70, 81, 83, 86–92, 95, 105–107, 109, 113, 114, 119, 121, 123, 125, 128, 132–134, 137–140, 142–144, 146, 148, 150, 152, 155, 156, 161–165, 173
Equilibrium constant, 12, 125, 132, 133, 135–138, 152, 155–157, 159–161, 164, 165
Excess enthalpy for a regular solution, 167, 189
Excess entropy for a regular solution, 167, 186, 189
Excess free energy for a regular solution, 185
Excess function, 167, 184
Exothermic, 20, 21, 29, 93, 137, 140, 143–145, 171, 172, 178
Expansion, 6–9, 11, 13, 14, 20, 35, 36, 49, 51, 52, 56–58, 115, 168
Extensive property, 6, 22, 27, 81, 82, 92, 184

F

Free energy change, 92, 93, 95–99, 116, 117, 125, 134, 135, 138, 142, 146, 150, 153–155, 157, 165, 177, 178, 189, 190, 198
Free energy change and its relation with activity
Free energy change and its relation with equilibrium constant, 134
Free energy change and its relation with enthalpy, 167, 173
Feasibility of a reaction, 81, 92, 93, 135
First law of thermodynamics, 15, 18–21, 27, 32, 35–37, 47, 48, 50, 58
Free energy, 1, 3, 4, 12, 81, 82, 91–93, 95–98, 105, 112, 116, 117, 125–129, 131, 135, 138, 140, 142, 144–146, 150, 153–155, 157, 165, 167, 177, 178, 184, 185, 189, 190, 198
Fugacity, 93, 94, 12, 125–131, 169

G

Generalized Carnot's cycle, 55
Gibbs-Duhem equation, 167, 172, 174–176, 187
Gibbs free energy, 3, 11, 82–84, 90–92, 99, 117, 167, 173, 174, 176
Gibbs free energy change and stability, 81
Gibbs-Helmholtz equations, 118, 122
Graphical method for determining partial molar properties, 182
Graphitization of carbon and entropy change, 47, 71

H

Hardening of steel and entropy change, 47, 72
Heat capacity, 15, 18, 21–23, 25–27, 31, 32, 34, 41, 70, 122, 123, 136
Heat capacity at constant pressure, 22, 23, 30, 34
Heat capacity at constant volume, 22, 59
Heat engine, 48, 50, 52, 54–56, 68, 82
Heat of formation, 29, 30, 179
Helmholtz free energy, 10, 81–84, 89, 90, 92, 99, 100, 111, 117, 173
Henrian activity, 188
Henrian limit, 186
Henrian standard state, 188, 189
Henry's law, 167, 186, 188–191
Hess's law, 32–34
Heterogeneous systems, 5
Homogeneous systems, 5
Hot mechanical working and entropy change, 71

I

Ideal gas and its enthalpy, 20, 28, 103
Ideal gas and its internal energy, 20, 24, 28, 36, 103
Ideal solution and enthalpy change, 180
Ideal solution and entropy change, 63, 180, 181, 185

Ideal solution and free energy change, 178, 184, 185
Ideal solution and its properties, 184
Integral molar properties, 175, 176
Intensive properties, 1, 6, 12, 22
Interaction between components in a solution, 167
Interaction coefficient, 189
Internal energy, 4, 10, 15–22, 24, 27, 28, 35, 36, 48, 50, 82, 83, 87, 88, 92, 100, 103, 117, 173
Internal energy vs. enthalpy, 35
Irreversible process and entropy change, 66
Irreversible processes, 1, 2, 7–9, 50, 51, 63, 65, 67, 68, 69, 87
Isobaric process, 6, 21, 35
Isochoric process, 6, 20, 88
Isolated systems and entropy change, 70
Isothermal process, 6, 8, 20, 28, 29, 33, 36, 49, 51, 113

K

Kirchhoff's Law, 15, 32, 34, 35

L

Latent heat of melting, 62, 72, 74, 76, 79, 109, 123, 146
Latent heat of vaporization, 62, 78, 108, 207–209
Limitations of Ellingham diagrams, 125
Liquid-vapor equilibrium, 108

M

Maximum work done, 83
Maxwell relations, 100, 102–104, 115, 116
Metallothermic reduction, 42, 43, 125, 147
Metal-oxide systems, 139, 142, 147
Metal-sulphide systems, 139, 151–153
Metastable equilibrium, 12
Mole fraction, 63, 138, 168–171, 174, 185, 186, 188, 189, 198
Multicomponent solution, 167, 177, 188, 189
Multicomponent system, 5

N

Natural processes and entropy change, 66, 67
Negative deviation form ideality, 167
Nernst's heat theorem, 112
Non-ideal solutions, 130

O

Oxidation, 3, 6, 30, 125, 139–141, 143–148, 150, 153–155, 160–162
Oxides of carbon, 147

P

Partial molar quantities, 167, 172, 173, 175, 176
Path functions, 1, 9, 10, 12, 16, 19, 22
Phase transformation, 3, 12, 15, 23, 30–33, 35, 47, 61–63, 71, 96, 105–107, 109, 125, 139, 145, 146, 153, 154
Polytropic process, 7, 13, 51
Positive deviation from ideality, 169, 171, 180
Pressure and its effect on equilibrium constant, 137
P-V-T relations, 7

R

Raoult's law, 167–170, 186–188, 190, 191
Raoult's law and negative deviation, 167, 169, 170, 187
Raoult's law and positive deviation, 167, 170, 187
Regular solution and enthalpy change, 167, 185, 189
Regular solutions, 167, 185, 186, 189, 191–193
Reversible adiabatic expansion, 51, 59
Reversible heating, 59, 60
Reversible isothermal expansion, 51, 58
Reversible process, 1, 2, 7–9, 13, 36, 48, 50–52, 56–58, 63, 68, 70, 82–84, 90, 91
Reversible process and entropy change, 57
Richard's rule, 109, 110

S

Second law of thermodynamics, 47–50, 55, 56, 60, 65, 83
Solid-liquid equilibrium, 106
Solid-solid equilibrium, 61, 106, 107
Solid-vapor equilibrium, 106, 109
Solute-solute interaction, 189
Spontaneous processes, 66, 68, 91, 93
Stabilities of metal oxides, 125, 144
Stabilities of metal sulphides, 150, 154

Standard state, 15, 29, 30, 35, 81, 94, 96, 99, 100, 126, 130–132, 134, 135, 142, 154, 155, 159, 167, 172, 175, 179, 188, 189, 197, 198
State properties, 6, 9, 12, 15, 27, 28, 56, 57, 69, 82, 172
Surrounding, 4, 5, 7, 12, 15, 16, 18–21, 28, 35, 48–51, 53, 57, 59, 60, 62, 70, 75, 86, 128, 172
Systems, 1–13, 15–21, 25, 27, 28, 30, 34–36, 47–51, 53–62, 64, 67–71, 75, 81–92, 95, 97, 100, 103, 105, 108, 127, 128, 133, 137, 138, 142–144, 147, 168, 169, 171, 173

T

Thermodynamics
- limitations, 16, 48
- usefulness, 2, 10

Third law of thermodynamics
- consequences, 81, 115
- significance, 117
- statement, 112–114
- verification, 81, 114

Trouton's rule, 81, 109, 110, 123

V

Van't Hoff Isochore, 135, 137